Light and Life Processes

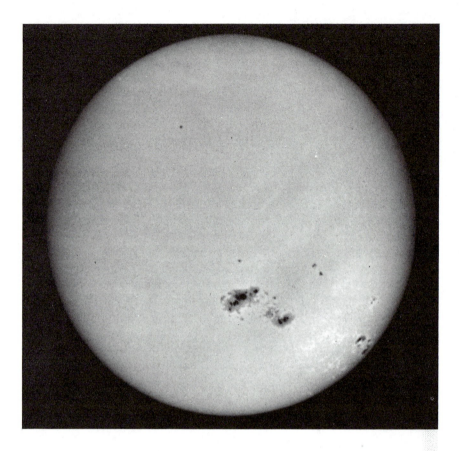

The Sun. Photograph taken with the 30-inch Thaw refractor at the Allegheny Observatory, Pittsburgh, Pennsylvania, showing sun spots. Courtesy of T. Reiland, University of Pittsburgh.

Light and Life Processes

Jerome J. Wolken

Carnegie-Mellon University
Pittsburgh, Pennsylvania

VNR VAN NOSTRAND REINHOLD COMPANY
New York

Copyright © 1986 by **Van Nostrand Reinhold Company Inc.**
Library of Congress Catalog Card Number: 86-7746
ISBN: 0-442-29348-8

Manufactured in the United States of America.

Published by Van Nostrand Reinhold Company Inc.
115 Fifth Avenue
New York, New York 10003

Van Nostrand Reinhold Company Limited
Molly Millars Lane
Wokingham, Berkshire RG11 2PY, England

Van Nostrand Reinhold
480 La Trobe Street
Melbourne, Victoria 3000, Australia

Macmillan of Canada
Division of Canada Publishing Corporation
164 Commander Boulevard
Agincourt, Ontario MIS 3C7, Canada

15 14 13 12 11 10 9 8 7 6 5 4 3 2 1

Library of Congress Cataloging in Publication Data
Wolken, Jerome J.
    Light and life processes.
    Includes index.
    1. Photobiology. I. Title.
QH515.W63 1986   574.19'153   86-7746
ISBN 0-442-29348-8

This book is dedicated to the memory
of my dear friend
<u>M.</u> E. Roux
who followed sun beams in search of
life's energy, photosynthesis

Contents

Preface

My purpose in this book is to explore with the reader the wondrous world of light that we live in. Radiation from the sun is a dominant force in our environment on Earth and it is of interest to examine the numerous ways light enters into life processes.

The solar spectrum that reaches the earth extends from the near ultraviolet to the infrared. Life evolved and functions within this narrow region of the electromagnetic spectrum of energy, the visible. Most photobiological phenomena center around the solar energy spectral peak in the blue-green. Ionizing far ultraviolet, and infrared radiation are detrimental to life. This book then is about the energetics of visible radiation to life.

The most important photoprocess to all life on earth is photosynthesis. For all animals with eyes, vision is paramount for seeing the world. Other photobehavioral responses are initiated by photoreceptor systems that detect and measure the kind of light in the environment. These result in oriented movement, growth, hormonal stimulation, sexual cycles, the timing of the flowering of plants, and many other extra-retinal responses that take place via the skin, nerves, and brain of animals. Light is essential to our health and useful as a tool for diagnosing disease and in preventive medicine.

The mechanisms that underlie these photoprocesses are highly complex and require knowledge of the physics of light, the chemistry and structures of the photoreceptor pigment molecules, the molecular structure of the photoreceptors in which the pigment molecules reside, and, finally, how the interaction of light with the photoreceptor systems produce the photobiological phenomena. Major technological advances in molecular photobiology have occurred, and photobiologists have greatly advanced our knowledge of photoprocesses in living organisms. I have indicated certain directions researchers have taken to understand these phenomena. Much is yet to be explored and understood in the world of light. We are at the beginning of the "Age of Light."

How light affects living organisms has been my life-long research interest, particularly the mechanisms of photosynthesis and vision. To observe the photobehavior of plants and animals in their natural environment, I have worked at various field and marine biological laboratories throughout the Americas and Europe. Some of these studies were collected in my previous books *Invertebrate Photoreceptors* and *Photoprocesses, Photoreceptors and Evolution*.

In this book I have tried to put together facts, observations, and interesting relationships between the various plants and animals, their photoreceptor pigment molecules, their photoreceptor structures, and their behavior. The examples I have chosen to illustrate the various photobiological phenomena are taken primarily from the organisms I have studied.

I have not directed this book to research scientists actively engaged in these various photobiological phenomena, for each is a highly specialized subject whose researches are published in scientific journals. Rather, I have addressed the book to students, scientists, engineers and educated lay persons curious about our photobiological world. The interested reader will seek further enlightenment in the references cited in the text.

Above all, I hope the book presents a glimpse of the wondrous world of light we live in and will awaken a greater interest toward appreciating the importance of visible light to life.

ACKNOWLEDGMENTS

I thank the publishers of my previous books: *Photobiology* (Reinhold Book Corporation, New York, 1968), *Euglena: An Experimental Organism for Biochemical and Biophysical Studies* (Appleton-Century-Crofts, New York, 1967), *Vision* (C.C. Thomas Publishers, Springfield, Illinois, 1966), *Invertebrate Photoreceptors* (Academic Press, New York, 1971), and *Photoprocesses, Photoreceptors and Evolution* (Academic Press, New York, 1975) for permission to reproduce certain figures and parts of the text. Permission to reproduce figures and data from my other books and papers published in various journals is gratefully acknowledged and is credited. Those who have sent me figures are acknowledged and much appreciated.

The Marine Biological Laboratory, Wood Hole, Massachusetts, library provided just the right environment to collect my thoughts in the summers of 1983, 1984, and 1985. The support I have received from the National Institutes of Health (NIH), National Science Foundation (NSF), National Aeronautical Space Administration (NASA) and especially from the Lions Eye Research Foundation of Pennsylvania over the years has permitted great freedom to pursue these photobiological studies, without whose financial assistance and interest this work could not have been accomplished.

For reviewing parts of the book and for their encouragement, my thanks go to Jonathan Wolken and Professor Mary Ann Mogus, and for their help in organizing the numerous rewrites, chapters and figures in the book, to Patricia C. Burns, Sue Caplan, and Paula Allen. Last, but not least, was the encouragement and help from my wife Tobey H. Wolken. For the many services that were extended to me during writing this book, my grateful thanks go to Mellon Institute, Carnegie-Mellon University, Pittsburgh, Pennsylvania.

JEROME J. WOLKEN

1 The Fascination of Light

Let there be light: and there was light.
Genesis 1:3

Life began as an inherent product of the sun and space.
Rutherford Platt, *The River of Life,* 1956

Light points the way—it is wondrous to experience the world of light from darkness. According to many early mythologies, all existence sprang from nothingness, out of darkness into the light. The contrast of light from darkness was so striking that ancient societies evolved legends describing the origin of light with the creation of life. Light became associated with the positive forces of life, goodness, purity, and holiness, while darkness came to represent the evil in life and the end of life itself. Living beings came from darkness before daylight and were thought to be luminous. The mystery of mysteries was the sun, whose majestic forces of light and heat were masters of Heaven and Earth, not only creating but sustaining life.

From our earliest records dating to the Babylonians about 6,000 years ago, light from the sun has been intimately related with the creation and evolution of life. In mythological accounts of the supernatural creation of life, the sun plays a central role. The records of ancient civilizations show that they were sun worshippers. We can read the sun's influence in early hieroglyphics and pyramids of the Greeks, Egyptians, and Persians, as well as the Mayan and Incan civilizations of the Americas. In Central and South America these early civilizations considered their rulers descendants of the primary celestial body and therefore "Sons of the Sun."

Light from the sun became a means to measure time. Clocks were devised to chart the passage of the sun and its changing position on the earth's horizon. Great efforts were undertaken to build pyramids to honor the sun and others were designed to analyze the motions of the heavens. The sun's movement became associated with the change of the seasons and with predictions of the solstices, important to agriculture.

The idea that light from the sun was the force that sustained life on Earth is associated with the beliefs of ancient Chinese, Egyptians, Babylonians, and Persians. They also believed that there are energy fields within the surrounding living organisms. The energy fields were thought to be blue, which was considered one of the higher colors of vibration. In ancient India the sun's true color was believed to be blue. There originated in Persia a mystical group of Hebraic origin known as the Kabbalists. Their first book of beliefs was written around 600 A.D. They flourished for many centuries, and by the fifteenth and sixteenth centuries amassed an extensive set of beliefs that linked light to creation and to life. The Kabbalists had recognized that light was a force that brought order and structure to living organisms. They believed that there is a blue energy permeating the universe and forming a field around all living beings and that, in addition to the direct light being absorbed, there is a reflected light that reascends through living bodies in a series of steps of light emissions. We now recognize this process as fluorescence. Needham (1956), a well-known biochemist and historian of the early history of science in China, wrote that the "correlative thinking" of the Kabbalists had great influence on scientific minds in the dawn of modern science. How remarkable that the science of today should have developed on the intuitions and myths of the past!

The beliefs of early mystics that light influences living organisms have given rise to a more current investigation as to whether there really is an energy field intimately associated with life. Let me cite only a few of these that bear on the question of light and life. Although not considered in the mainstream of scientific research, they remain more than a curiosity. In the 1920s A. G. Gurwitsch proposed the concept of *mitogenetic radiation*. He was convinced that ultraviolet light was involved in the basic processes of life. In the 1930s claims were made that there was a blue, near-ultraviolet light emission from growing and dividing cells (Gurwitsch, 1968). Reinvestigation of this phenomenon failed to confirm Gurwitsch's observations, but interest emerged again in the 1970s from new experiments that seemed to confirm these earlier observations. In investigations of dividing amoeba, bacteria, yeast, germinating seeds, and animal tissue cells, a weak near-ultraviolet light emission, a luminescence, was observed (Slawinska and Slawinski, 1983; Quickenden and Que Hee, 1981; Quickenden, Comarond, and Tilbury, 1985).

Related in some way to the research for energy fields associated with living organisms are the series of studies of Wilhelm Reich in the 1940s. Reich's mission was to find the "life energy" or, as he called it, the *orgone energy*. He observed in his experiments that the emitted light was bluish, and when the emitted light was measured it was found to approximate the spectrum of blue light (Reich, 1948).

In this context another interesting related development occurred during the 1930s in the research of Kirlian. He developed a unique method observing energy fields of living organisms, known as Kirlian photography (Kirlian and Kirlian, 1961). Kirlian photography became a popular subject for re-investigation in the 1970s when his methods were rediscovered. The Kirlian process employs a form of contact photography, performed with a high-voltage alternating current between two condenser plates. The process is associated with an electrical discharge, better known as *corona* discharge or radiation-field photography. These discharges appear as a bluish flame of light, and the resulting photographs are startling. Plant leaves, animals, and even the human body and appendages, hands, arms, and feet, all show pulsating multicolored lights. The color of light associated with the images of living organisms is predominantly blue and is believed by those using Kirlians method to indicate the state of health or degree of well-being (the aura) of the individual plant or animal.

Even if Gurwitsch's mitogenetic radiation, Reich's orgone energy, and Kirlian photographic images do not measure the energy fields of living organisms, photobiologists have shown that near-ultraviolet blue light is very much associated with life. It may well be that the origin and evolution of life are directly connected to the shorter wavelengths of light, the near-ultraviolet, which through a long series of hereditary mutations are the basis for natural selection and evolution.

It is important from an evolutionary point of view that near-ultraviolet blue light affects the behavior of lower plants and animals, and that the spectral sensitivity peak for vision in most insects and birds is in the near-ultraviolet. But the solar spectrum extends beyond the ultraviolet through blue, green, yellow, orange, and red, from 300 nm to 1000 nm. So, cells synthesize pigment molecules for reception of these energies, the very energies that power the key photoprocesses of photosynthesis, phototactic movement, growth and flowering of plants, reproductive cycles, and vision in animals.

Therefore, solar radiation was essential for the development of life on Earth, and in the absence of light, life as we know it would not have been possible. Our purpose then is to examine the world of light we live in by seeing how light affects life on Earth. More specifically, the focus of this book is in the interaction of light with the biophysics of "seeing." Toward this end we must understand the physical nature of light, the chemistry of the photoreceptor pigment molecules, the molecular structure of the photoreceptors in which these pigments reside, and the photochemistry that happens inside them to result in the various photobiological phenomena. The deeper we delve into the forces that sustain life on earth, the more clearly life emerges as a light-mediated phenomenon.

2

The Source: The Sun and Radiation

Now consider the excellence of the Sun, prime prince
and controller of the world, favoring and forwarding
every life that is. . . .
Jean Fernel, 1542.

. . . the whole of the energy which animates living
beings, the whole of the energy which constitutes life,
comes from the sun.
S. Leduc, *The Mechanisms of Life,* 1911.

THE SUN

The sun is the primary energy source for Earth and supplies us with light as well as heat. The sun is a star that occupies a very important role in our planetary system; it is the center of our solar system around which the planets orbit.

What makes the sun so important to all life on Earth? To answer this question we must know something of its origin, composition, properties, energetics, and the spectrum of energy that strikes the earth. With this information we can begin to see how the sun's energy interacts with the molecules of living organisms to affect the behavior of life on Earth.

It is estimated that the sun was formed about 4.5-5 billion years ago. The sun is too hot to be solid and is comprised of gaseous elements, primarily hydrogen and helium. All of the naturally occurring elements found on earth have been detected in the sun, though only in minute quantities.

The sun is relatively large compared to the earth; its diameter is 169 times that of the earth and its surface area is 12,000 times that of the earth. The sum of all the planets, moons, and meteors in our solar system account for only 2% of the mass in the solar system, meaning that the sun contains 98% of

the total mass. Because of its mass the sun has a powerful gravitational pull; at its surface its gravitational force is 28 times stronger than that of the Earth.

The sun has changed little during the last 4.5 billion years and is expected to remain relatively unchanged for a few more billion years. Afterward, the sun may begin to transform into a "red giant," as it grows to 100 times its present diameter, and then begin to collapse into a "white dwarf" with a diameter one-half that of the earth.

The sun produces immense amounts of energy through thermonuclear fusion, a process by which small atoms like hydrogen fuse to form larger atoms in the synthesis of helium. When this fusion happens, the combined masses of the fusing atoms are more than the mass of the atom they form, and some of this extra mass is converted into energy. The equivalence of mass and energy is well known through Einstein's relationship, $E = mc^2$, in which c is the speed of light. Only about 0.7% of the sun's mass loss is actually converted into energy, but because the sun loses 5.6×10^{16} kg of matter a day its energy output each second is equivalent to billions of our largest hydrogen bombs.

Solar flares are powerful explosions on the sun that hurl matter into space and shower the earth with high-energy radiation and atomic particles. Estimates of this energy in terms of hydrogen bomb equivalents range from a few billion to ten trillion megatons. Terrestrial consequences of solar flares have been associated with magnetic storms and auroral displays. It is estimated that a large flare releases sufficient energy to fulfill the energy requirements of the United States for thousands of centuries to come. Almost all this energy is converted into photons.

The biosphere harvests about 1% of sunlight that is incident upon the earth. Even though the earth receives only a small fraction of the sun's total energy output (about 1 part of a trillion), that amount is still enormous. On any given day, the amount of solar radiation reaching the earth is roughly equal to the sum of all energy stored since the beginning of the earth in fossil fuels and the heat stored in the ocean waters combined. The earth retains only a small portion of the energy incident upon it; terrestrial radiation is mainly of infrared rays, with a maximum near 11μm, and is re-radiated out into space in the form of infrared radiation.

In studying the radiation emitted by the sun, it is found that the sun acts very much like a perfect absorber and transmitter of radiation, or a blackbody. The intensity of radiation emitted by a blackbody, for any given wavelength, depends wholly upon its temperature. The sun has an average surface temperature of 5800°K. The variations from this ideal are mostly caused by the variations in the sun's temperature as one moves away from the center. In Figure 2.1 we can see that the radiation emitted most frequently lies in the visible range, with some radiation falling into the ultraviolet and infrared region of the spectrum (Fig. 2.4).

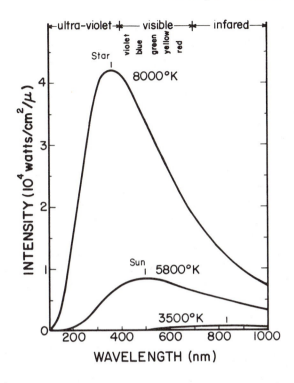

Figure 2.1 Blackbody curves showing how the intensity of radiation from perfectly efficient light emitters varies with the wavelength of light. The curves give an approximation of how the intensity of their radiation varies with color. An 8000°K star (such as Altair) is seen to emit most of its energy in the visible ultraviolet, while a 3500°K star (such as Antares) emits most of its energy in the invisible infrared. The Sun, a 5800°K star, emits most of its energy in the visible range. The intensity emitted per unit area of the star varies enormously with temperature.

THE PHYSICAL NATURE OF LIGHT

What exactly is light? This question has been a major puzzle ever since the advent of modern physics, and explanations before 1900 almost invariably fell into two categories: light is either a wave or a particle. Because it exhibited characteristics in favor of both arguments, there were always problems with either position. It was not until this century that a seemingly simple answer was proposed: light is *both* a wave and a particle.

The wave nature of light is understood through the model of electromagnetic radiation. Electromagnetic radiation is composed of an electric and magnetic field traveling through space with a constant velocity. The two fields are perpendicular to each other and the direction of propagation of the wave is perpendicular to each of the fields. Because a changing electric field will create a changing magnetic field and a changing magnetic field will create a changing electric field, such constantly changing fields lead to the reinforcement of one field by the other. This reinforcement allows the electric and magnetic fields to regenerate each other for an indefinite length of time if left to themselves. This method of propagation is how electromagnetic energies in space, such as light from distant stars, can travel vast distances to us.

Electromagnetic radiation can be made to form interference patterns. This phenomenon, found only with waves, points toward the wave character of light. Because electromagnetic radiation is a transverse wave it may also be polarized. Both interference and polarization are phenomena useful in studying the behavior of living organisms.

Light is characterized on the basis of its wavelength or its frequency. The wavelength, λ, is the distance from one identical point on consecutive cycles to another, for example, from peak-to-peak or trough-to-trough (Fig. 2.2). The frequency v is the number of cycles the wave goes through in one second. In other words, if the wave goes from peak to trough to peak again five times in 1 sec, then its frequency is five per sec, or 5 H. If both the frequency and wavelength are known, then the velocity of the wave can be easily calculated. Because light is a traveling wave, you would see it move in space if you could possibly fix your eyes on one peak; if you could fix your eyes on a single, immobile point in space, you would see the wave oscillate from peak to trough to peak, and so on.

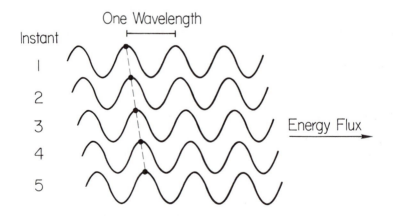

Figure 2.2   Electromagnetic radiation as a wave motion. The same wave is shown at five successive instants of time. The dashed line connecting points of similar amplitude on the wave shows the wave moving to the right as time passes.

Because the frequency gives the number of peaks that pass a single point in one second, and the wavelength gives the distance between two consecutive peaks, the distance from the first peak passing through that point, at time = 0, and the peak passing through at time = 1 sec, is simply the distance between two peaks times the frequency or λv. In this manner we find the velocity, which is the distance traveled per second by a point on the radiation wave. For electromagnetic radiation this velocity is constant in vacuum no matter how long or short the wavelength is, and is found to be 3×10^8 m/sec, which is the speed of light *(c)*. So for electromagnetic radiation $\lambda v = c$.

The particle theory of light is understood through modern quantum theory. According to the theory, light is transported in wavelike bundles of energy called photons or quanta. The quantum and wave properties of radiation are not two separate qualities that together make light; the two are intimately related. Max Planck (1922) discovered the direct relationship between frequency of electromagnetic radiation and the energy of its quanta. Albert Einstein (1905) extended Planck's relationship to include light. Einstein had shown—and this feature was part of Planck's derivation of the blackbody spectrum as well—that the energy of a light quantum was proportional to its frequency. That is, each photon has the energy $E = hc/\lambda$, where h is Planck's constant (6.62559×10^{-27} erg·sec), c is the velocity of light and λ is the wavelength of the light. Then the energy of a single quantum can be calculated from $E = hv$, because $v = c/\lambda$. This equation shows that the higher the frequency of the radiation, the greater the energy. For example, quanta of violet light (8×10^{14} per sec) would be more energetic than quanta of red light (4×10^{14} per sec), and X-rays would be even more energetic because their frequencies are higher than any of the visible frequencies. Therefore, frequency is inversely proportional to wavelength and the shorter the wavelength, the greater the energy.

Einstein postulated that all the energy of a single light quantum, or photon, is transferred to a single electron. This one-to-one relationship between a light quantum and a particle of matter is of key importance in photochemistry. The principle that one quantum of light can bring about a direct primary photochemical change in exactly one molecule of matter is known as Einstein's Law of Photochemistry. In other words, a photon with sufficient energy may strike an electron in an atom and change the chemical properties of that atom.

THE ABSORPTION OF LIGHT

The ability of a substance to absorb light is determined by its atomic structure, that is, by the arrangement of electrons in different orbitals about the nucleus of each atom. Those electrons nearest the nucleus have relatively low energy, while those in orbitals farthest away from the nucleus have a higher energy. To move an electron from an inner orbital to an outer orbital requires energy. When photons of light strike a molecule that can absorb the light, an electron in one of the orbitals may absorb the photon and gain energy sufficient to move it farther away to an outer orbital of a higher energy level. The molecule is then referred to as being in an *excited state.*

In some molecules in the excited state the high energy electrons do not escape from the molecules but return to their original low energy orbitals again, and the molecule is said to return to the *ground state.* Some of the absorbed energy may reappear as light, as in fluorescence, when the electron returns to the ground state. There are four types of processes that a molecule in the excited state may undergo: (1) emission of light or a radiative transition, (2) a radiationless transition between two states without chemical reaction, (3) electron excitation energy transfer, and (4) chemical reaction.

For molecules with an even number of electrons, in practically all cases, the photochemical behavior is describable in terms of singlet ($S = 0$) and triplet ($S = 1$) excited states. The distinction between single and triplet excited states of molecules that absorb light is of great importance in the photochemistry of pigment molecules and therefore in photobiology. The absorption of light by a biological system and the emission of light from excited molecules are quantum phenomena. Therefore a proper description of the light either of absorbed or emitted during a reaction should contain the number of photons per second per unit wavelength.

THE ELECTROMAGNETIC SPECTRUM

Newton in 1666 demonstrated that sunlight is not just white light and showed that if you shine light through a glass prism and pass it on to a screen you will see a continuum of colors: red, orange, yellow, green, blue, and violet. These colors are frequencies or wavelengths of the visible spectrum of electromagnetic radiation.

The full spectrum of electromagnetic radiation extends from γ and X-rays less than 0.01 Å long, through ultraviolet, visible, and infrared radiation, to radio and electric waves that are kilometers long (Fig. 2.3). Abundant data is now available on the spectral distribution and intensity of solar radiation incident on the surface of the earth; this radiation is constantly being measured. The sun's spectrum of energy that strikes the earth extends from the near-ultraviolet to beyond the red to the infrared. But the visible part of the electromagnetic spectrum is a very narrow region from about 390 nm to about 760 nm (Fig. 2.4).

The radiation energy effective for photobiology lies between 300 nm and 900 nm. Practically all photobiological behavior of plants and animals—

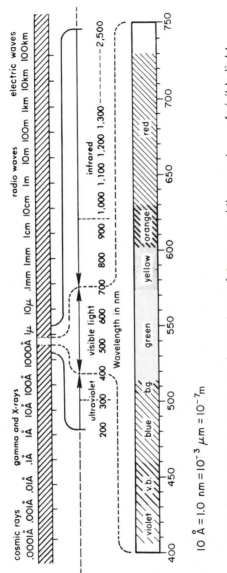

10 Å = 1.0 nm = 10^{-3} μm = 10^{-7}m

Figure 2.3　The electromagnetic spectrum of energy and the spectrum of visible light.

11

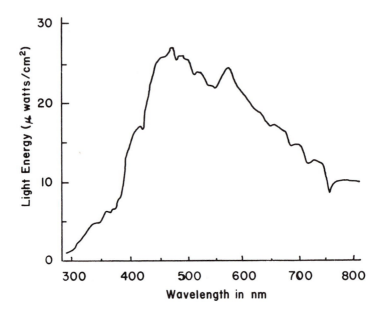

Figure 2.4   Solar spectrum that strikes the earth on 82 cm² taken between 1:00 P.M. and 2:00 P.M. in Cadarache, France. (Courtesy of Dr. P. Guerin de Montgareuil, Atomic Energy Commission, France.)

photosynthesis, phototropism, phototaxis, photoperiodism, and vision—utilize this range of radiation energy for excitation. The spectrum of solar radiation that reaches the surface of the earth covers this photobiological range, with a maximum around 500 nm about which these phenomena cluster.

Ultraviolet radiation from 200 nm to 300 nm is absorbed by proteins and nucleic acids; such absorption produces damaging effects on cells and greatly increases the frequency of mutations. The wavelengths of ultraviolet radiation that produce mutations correspond to the absorption spectrum of the genetic molecule, DNA. Evolving organisms were able to repair these damaging effects through a phenomenon known as photoreactivation via visible light. Photoreactivation continues to be effective in bacteria, fungi, plants, and animals.

Near-ultraviolet light from about 300 nm to 400 nm and blue light from 400 nm to 500 nm are of considerable interest in photobiological phenomena. Action spectra for phototropism, phototaxis, and the spectral sensitivity for vision of most insects show a major response near 360 nm. Because action spectra are indicative of the absorption spectrum of the molecules involved in a process, we infer these organisms possess photoreceptor molecules to absorb this energy.

Radiation in the red part of the spectrum from 600 nm to 700 nm is important for chlorophyll synthesis and photosynthesis. Radiation from 660 nm and into the near-infrared is important for plant and animal growth, the timing of plant flowering, sexual cycles in animals, and pigment migration. The timing of flowering cycles in plants is controlled in the red, near part of the spectrum by shifting of light between 660 nm and 730 nm. Bacterial photosynthesis takes place even further in the red, near 900 nm. Infrared radiation beyond 900 nm is mostly absorbed by atmospheric water vapor and by water that surrounds living cells.

3   *The Environment for Life*

> The growth of physical science has provided the
> speculative biologist with a very accurate and extensive
> description of the physio-chemical structures of the
> material universe and with a well founded confidence in
> his right to make use of the description in investigating
> the relationship between life and the environment.
> Laurence J. Henderson, *Fitness of*
> *The Environment*, 1913.

THE ENVIRONMENT OF EARTH

The planet Earth is a small part of the Milky Way, which in turn is but one of the myriad of known galaxies. What were the environmental conditions on the primordial Earth? When did Earth form a crust and what was its chemical composition? When did the planet cool enough for the oceans to condense? What was the early atmospheric composition, and how has the atmosphere changed? What part has solar radiation played in chemical synthesis involved in the origin and evolution of life? In search of answers to these questions we are seeking clues to solve the mystery of what made life possible on Earth. This challenge and the related challenge to discover whether there is life elsewhere in the universe is bringing new excitement to astronomy, chemistry, physics, geology, paleontology, and molecular biology.

Life as we recognize it on Earth has not yet been detected on other planets in our planetary system. However, from satellite explorations of the moon, Jupiter, Mars, and Venus and from chemical analysis of the atmosphere, planetary dust, and recovered meteorites, much is being learned about the atmospheric gases and chemistry of extraterrestrial environments. The formation of complex organic molecules has been found in the atmosphere of

Jupiter and in comets and meteorites. Additional evidence is accumulating to indicate the universality of certain organic molecules that we associate with life.

From such data, extrapolation back to the early history of the Earth's environment is being made. The Earth may be unique in that its atmosphere, temperature, and chemical and physical environmental conditions were suitable for the development and continuance of life.

Our knowledge of the history of the Earth's formation and structure comes from geology, geophysics, and geochemistry. It is estimated that the Earth was formed 5 to 6 billion years ago and hardened into its present shell. This information is based on evidence from radioactivity that the oldest reliably dated terrestrial rocks have ages in excess of 3 billion years. Assuming that the present abundance of ^{207}Pb resulted entirely from radioactive decay of ^{235}U, it turns out that Earth's age lies between 3.2 and 5.7 billion years. The estimated age of the crust of Earth therefore is around 4.5 billion years.

Although there is general agreement on its age, the question of the origin of Earth and the other planets remains unanswered. Some hold that Earth gradually formed from hot gases thrown off by the sun; others believe that meteorites gradually aggregated to form one large solid planet. Whichever theory is presently subscribed to for the evolution of the planets, Earth and the sun are believed to have a common origin.

Precisely when life arose on Earth is not known, but it appears to have happened very soon after it was formed some 4 billion years ago. Paleontologists are piecing together the past history of life on Earth by carefully exploring the preserved fossil morphology. Modern techniques for dating have given us some idea of the age of fossilized remains and the time when life emerged. A variety of life forms evolved with time and we can follow their progress in the geological strata of Earth (Fig. 3.1).

A detailed study of the rocks that make up the surface layers of Earth enables us to trace small changes in an ever-changing sequence of different life forms. Almost all preCambrian rocks are severely folded, and the folding has blotted out most of the relics of life. We have fairly definite traces of the existence of calcareous seaweeds, protozoa, and marine worms, but there is no evidence of life on land. Probably the intense folding confined the seas into smaller areas than at present, and the land surface of the globe was largely covered by rainless deserts. There were, however, at least two ice ages before the Cambrian. The evidence of terrestrial rocks shows that Earth was already over a billion years old in Cambrian times.

Preserved in the Cambrian rocks are the remains of most of the main animal phyla. The only important phylum missing is that of the vertebrates (phylum Chordata). The vertebrates did not appear until the Silurian, though presumably their boneless, wormlike ancestors were present in the Cam-

Figure 3.1   Geological time scale in relation to evolved life forms.

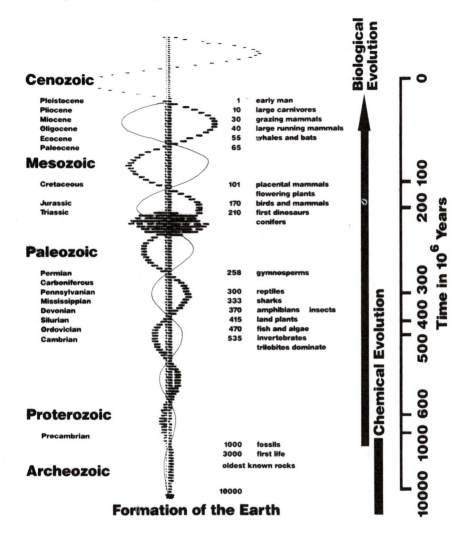

Cenozoic		
Pleistocene	1	early man
Pliocene	10	large carnivores
Miocene	30	grazing mammals
Oligocene	40	large running mammals
Ecocene	55	whales and bats
Paleocene	65	
Mesozoic		
Cretaceous	101	placental mammals
		flowering plants
Jurassic	170	birds and mammals
Triassic	210	first dinosaurs
		conifers
Paleozoic		
Permian	258	gymnosperms
Carboniferous		
Pennsylvanian	300	reptiles
Mississippian	333	sharks
Devonian	370	amphibians insects
Silurian	415	land plants
Ordovician	470	fish and algae
Cambrian	535	invertebrates
		trilobites dominate
Proterozoic		
Precambrian		
	1000	fossils
	3000	first life
		oldest known rocks
Archeozoic		
	10000	

Formation of the Earth

Biological Evolution

Chemical Evolution

Time in 10^6 Years

0
100
200
300
400
500
600
1000
10000

brian sea. In the upper Silurian there were already fish with fins. These fish rapidly developed paired fins from lateral folds. In the Devonian they developed bones and probably about the end of that period left the water. Most of the main animal types were thus already differentiated in their comparative anatomy, embryology, and biochemistry.

THE CHANGING ATMOSPHERE

What was the early atmosphere's composition and how has the atmosphere changed as life evolved? The early atmosphere of Earth must have been totally different from the present nitrogen-oxygen mixture existing today (Table 3.1); it had virtually none of the free oxygen it has today (Schopf, 1983). Instead, it is thought that the atmosphere was reducing; that is, hydrogen was the most prevalent gas. If oxygen were present, it would have immediately combined with various materials that existed then in a chemically reduced state, as well as free iron, sulfur, and other elements.

Oxygen was probably first introduced into the atmosphere by the photodissociation of water vapor exposed to ultraviolet radiation. Hydrogen would also have been produced, but would have escaped comparatively quickly from Earth's atmosphere. At first, the oxygen would also have been rapidly

Table 3.1   Composition of Air (Dry, Clean Air Near Sea Level)

Constituent Gas and Formula	Content (% by Volume)
Nitrogen (N_2)	78.084
Oxygen (O_2)	20.9476
Argon (Ar)	0.934
Carbon dioxide (CO_2)	0.0314
Neon (Ne)	0.001818
Helium (He)	0.000524
Methane (CH_4)	0.0002
Krypton (Kr)	0.000114
Hydrogen (H_2)	0.00005
Nitrous oxide (N_2O)	0.00005
Xenon (Xe)	0.0000087
Sulfur dioxide (SO_2)	0 to 0.0001
Ozone (O_3)	0 to 0.000007 (summer)
	0 to 0.000002 (winter)
Nitrogen dioxide (NO_2)	0 to 0.000002
Iodine (I_2)	0 to 0.000001
Ammonia (NH_3)	0 to trace
Carbon monoxide (CO)	0 to trace

Sources:U.S. Standard Atmosphere, 1962, p. 9; National Aeronautics and Space Administration; United States Air Force; and United States Weather Bureau, Washington, D.C.

removed as it reacted with the materials of the earth's crust, but as they became fully oxidized, the level of free oxygen in the atmosphere would have started to rise. Urey (1952) has pointed out, however, that the oxygen level in the atmosphere would not have risen to anything like today's near 20% as a result of photodissociation alone because as free oxygen accumulates it tends to shield Earth from the ultraviolet radiation required to bring about photodissociation. Thus the process is a self-regulating one.

There is no doubt that a reducing atmosphere was present at the stage in Earth's history when life first arose. This certainly may seem unlikely in view of the fact that almost all living organisms today are dependent on oxygen for respiration. But as Oparin (1938) pointed out, the presence of a reducing atmosphere during the early stages of life was, paradoxically, a necessity. The transition from mainly anaerobic to mainly aerobic conditions on Earth was brought about not only by geochemical and photochemical processes but by life itself through photosynthesis.

Earth's atmosphere has several well-defined layers (Fig. 3.2). The lowest and most dense layer is the troposphere. This region extends up about 15 km, throughout which the relative proportions of the major gases remain constant due to thorough mixing by convection currents. Between 15 km and 80 km above the surface, a number of distinct layers are present, among which

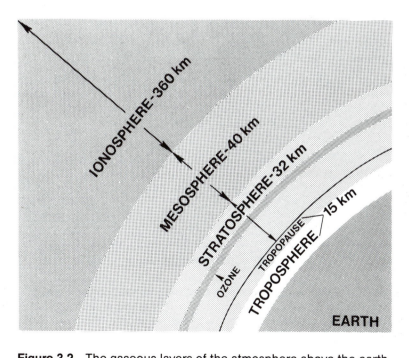

Figure 3.2 The gaseous layers of the atmosphere above the earth.

there is very little vertical circulation. This region of the atmosphere is called the stratosphere. Throughout the troposphere and the stratosphere the predominant molecular forms are N_2 and O_2. The region above the stratosphere, from 80 km to 100 km, is rich in ionic forms and is thus named the ionosphere. The presence of these ions suggests that either high temperatures prevail or that photochemical reactions and ultraviolet photoionizations are taking place in this region.

At present we can divide the gases that make up the lower atmosphere into two groups. The first group, nitrogen and oxygen, makes up practically 100% of the atmosphere near Earth's surface. The other group includes water and rare gases. Water occurs in the air in the form of water vapor and can be present in concentrations of up to 2-3% by volume. Dry air always contains ozone and radon. The concentration of radon, a product of the disintegration of radioactive elements in Earth's crust, is minimal. The atmosphere normally contains 10^{-6}% ozone (O_3), the concentration increasing with the height above Earth's surface. The data on atmospheric gases are recorded in Table 3.1.

Most of the ozone resides in a thin layer located about 24 km above sea level. The ozone in conjunction with water, O_2, and CO_2 effectively absorbs almost all radiation below 300 nm (Fig. 3.3).

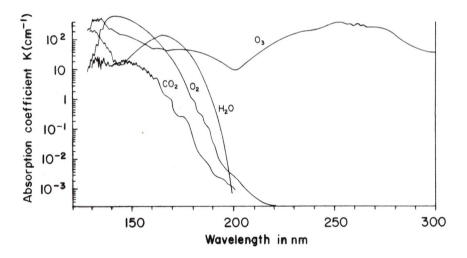

Figure 3.3 Ultraviolet absorption spectrum in atmospheric gases. *(From Wolken, 1975. Spectra taken in part from L.V. Berkner and L. K. Marshall, 1964; and ozone absorption spectrum from K. Watanabe, 1959, and E. Vigroux, 1969.)*

When an O_3 molecule is hit by a high energy photon, it splits into an oxygen molecule and an oxygen atom:

$$O_3 + photon \rightarrow O_2 + O.$$

But at altitudes of less than 27 km the O_2 and O created by the process react almost immediately to form a new molecule of ozone. The rates of ozone dissociation and the combination of O_3 and O are roughly equal, so the small amount of ozone in our atmosphere remains constant, although in a state of flux.

Ultraviolet radiation more powerful than this level, with wavelengths shorter than 100 nm, is absorbed by almost all atmospheric gases. Absorption usually takes place at altitudes of over 182 km. This absorption, coupled with the fact that the sun emits relatively few extremely energetic rays due to its temperature and blackbody spectrum, assures that few of these rays penetrate the atmosphere.

If the biosphere is the major source of the molecular oxygen in the present atmosphere, then it is likely that before the biosphere came into being there was very little molecular O_2 in the atmosphere. This possibility implies in turn that the ozone screen might have been absent from the prebiological atmosphere, thus allowing the penetration of considerable amounts of solar short-wavelength ultraviolet radiation deep into the atmosphere. Radiation of wavelengths as short as 200 nm could conceivably have reached Earth's surface. Thus the primitive atmosphere would have been abundantly supplied with a free energy source for photochemical reactions.

These arguments require modification if there were other sources (besides the biosphere) contributing molecular oxygen to the atmosphere. For example, recent NASA film from a surface ultraviolet camera spectrograph left on the moon by *Apollo 16* indicate that Earth's present source of oxygen may not be primarily photosynthesis, but that oxygen is formed by solar ultraviolet radiation as a result of the dissociation of water vapor in the upper atmosphere.

As a result of biological and geological processes, a variety of gases are emitted from Earth's surface into the atmosphere. The vast majority of the trace gases emitted into the atmosphere are in reduced states, such as hydrogen sulfide, ammonia, and methane. By contrast, materials returned to the surface from the atmosphere, usually by dissolution in raindrops or by dry deposit, are highly oxidized, such as sulfuric acid, nitric acid, and carbon dioxide. The link between these reduced and oxidized forms is supplied by atmospheric photochemical reactions, thereby giving rise to a chemical cycle whereby reduced gases are emitted into the atmosphere, photochemically oxidized, and then removed from the atmosphere.

That reduced atmospheric gases are oxidized is not surprising; the large

atmospheric abundance of oxygen thermodynamically favors their oxidation. However, molecular oxygen rarely is directly responsible for oxidizing these reduced gases. The oxygen-oxygen bond is relatively strong (120 kcal/mole) and thus molecular oxygen does not react readily with most reduced gases at atmospheric pressures and temperatures.

TEMPERATURE

The temperature of Earth's crust would for a long time have been well above 100°C. Therefore all the water must have been present in the atmosphere as vapor until Earth cooled sufficiently to allow water to condense. The amount of hydrogen in the atmosphere must have gradually decreased as Earth cooled. There is overwhelming geological evidence that for at least 2 billion years the range of temperature on Earth's surface has been within the limits now experienced, that is, from slightly below the freezing point of water to a little above 23.5°C. If this continuity had not been the case, life probably could not have developed and continued on Earth.

LIGHT IN THE SEA

The cycle of life in the sea, like that on land, is fueled by the Sun's visible light.

The ocean and the atmosphere constitute a single system that functions as a chemical plant. About 70% of the planet Earth is covered by water, and it is said that "Earth" should really be called "Ocean."

Table 3.2　Composition of Seawater[a,b,c]

Elements	Ions	Content (% by weight)
Sodium	Na^+	1.06
Magnesium	Mg^{2+}	0.130
Calcium	Ca^{2+}	0.040
Potassium	K^+	0.0388
Chlorine	Cl^-	1.90
Sulfate	SO_4^{2-}	0.267
Carbonate	Co_3^{2-}	0.007
Bromine	Br^-	0.0065
Borate	H_3BO_3	0.0026
Other elements		0.09

[a]The large number of elements known to occur in seawater indicates that probably all of Earth's naturally occurring elements exist in the sea.

[b]Seawater is by weight 96.46% water, of which oxygen makes up 85.79% and hydrogen 10.67%.

[c]Refer to Culkin (1971) for additional data.

Sea water is another nearly constant characteristic of our environment. The salts of the open oceans rarely exceed 3.8% or fall below 3.3%, and the average is about 3.45%. Nine ions constitute 99.5% of the salts in solution (Table 3.2) and are found in constant proportions to one another.

The remarkable consistency of the chemical composition of sea water is of great biological importance, especially because many scientists believe that life made its appearance rather early in the 4.5 billion years of earth history—most likely in the form of a bacterial cell. If we compare the relative proportions of the different ions present in the blood serums of various animals, we find a great similarity. Macullum (1926) suggested that the circulating fluids of all animals originally came from the primeval seas.

In summary we see that water absorbs light, transforming radiant energy into heat. Both ends of the visible spectrum, especially the red end, are selectively absorbed. And as one descends into the ocean the appearance of daylight changes with increasing depth from white, to blue, to blue-green (Fig. 3.4). We know that nitrogen is the dominant gas in our present atmo-

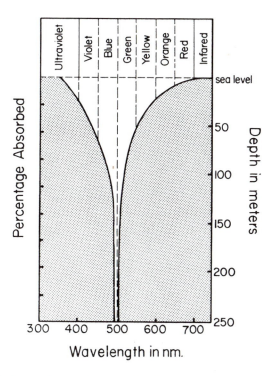

Figure 3.4 Solar irradiation in water. Highest energy wavelengths are found in the blue-green.

sphere (Table 3.1), and that it might have held a similarly prominent position in Earth's primitive atmosphere as well. This dominance allowed large amounts of blue light to strike Earth's surface (Fig. 2.4) and to penetrate deeply into the oceans (Fig. 3.4). So it is not surprising that all life has evolved with a special sensitivity to blue light; it is, after all, the dominant waveband in our environment.

4

The Molecules We Associate with Life

The molecules out of which living material is made
contain large stores of internal energy. . . . And from
what source do the molecules of living creatures here
on Earth get their internal supplies of energy? . . . Plants
get it from sunlight, and animals get it from plants, or
from other animals . . . So in the last analysis the energy
always comes from the Sun.

Fred Hoyle, *The Black Cloud,* 1967.

Life is dependent on the interaction of physical forces with living matter and chemical reactions within living cells. With the information we have on solar radiation, the spectrum of electromagnetic energy that strikes Earth, and the environment, we must now inquire what are the major molecules, the building blocks, that constitute living organisms. The story of life on Earth is the story of carbon chemistry. The carbon atom with its four electrons can form a myriad of chains of varying length and shape—truly a versatile building block. The great number of its possible variations explains why carbon compounds are so common on Earth and play such a central role in all living matter. Let us then identify these molecules, describe their structure, and briefly consider the properties that enable them to function in living organisms.

The main substances of living cells are water, certain inorganic mineral salts, and a variety of organic compounds. The organic compounds—the proteins, lipids, carbohydrates, and nucleic acids (DNA, RNA)—for the most part are considered the essential molecules of life. These molecules are structured of surprisingly few atoms: carbon, hydrogen, oxygen and nitrogen, which together constitute 99% of living matter. For example, the human body is about 15% proteins, 15% lipids or fats, about 1% carbohydrates, 5% inorganic materials, and the remainder water.

Let us briefly examine these organic molecules, beginning with proteins, for it will be necessary to refer to these molecules and their structures in our discussion of photoprocesses and photobiological phenomena.

PROTEINS

Proteins are of special interest not only because of their size and complexity but also because of their variety, versatility, and species-specificity. No two species of organisms possess exactly the same proteins.

Proteins consist of thousands of atoms, mostly hydrogen, oxygen, nitrogen and carbon. The problem of how these atoms are arranged in the protein molecule is very intriguing and challenging to researchers in the physical and biological sciences. Proteins are the largest and most complex molecules known to us. They vary in molecular weight from the order of thousands to the order of millions. There are tens of thousands, perhaps as many as 100,000, different kinds of protein in a single human body.

Proteins serve a multitude of purposes, some of which we are just now discovering. By and large they function to regulate metabolic processes as catalysts for biological reactions, and they comprise the "mortar and bricks" of which most cellular structures are made.

Proteins are structured of long chains of about 20 different amino-acid residues (Fig. 4.1). An amino-acid residue is the group of atoms that remains after a molecule of water has been removed from an amino acid. Long chains of amino-acid residue are called polypeptide chains. The chains are usually very large. For example, in ovalbumin, the principal protein of egg white, about 400 amino-acid residues form a single polypeptide chain. The kind and number of different amino-acid residues in a protein molecule can be determined by chemical analysis of the protein.

Proteins are generally characterized as either the globular or the filamentous. The filamentous are composed of elongated chains with a considerable amount of folding of the chain. In the study of the structure of a protein there are two questions to be answered: 1) What is the sequence of amino acids in the polypeptide chain? and 2) What is the way in which the polypeptide chain is folded in the space occupied by the molecule? X-ray studies of native proteins and synthetic polypeptides led Linus Pauling (1960) to propose a structure of greatest stability, the α-helix. The α-helix has a spiral chain of repeating amino acids held together by hydrogen bonds (Fig. 4.2). The helix contains about four amino acid residues for each full turn of the spiral. The α-helix is a one-dimensional subcrystalline arrangement. Another type of structure is the β-configuration or pleated sheet (Pauling, 1960). Here two or

AMINO ACIDS

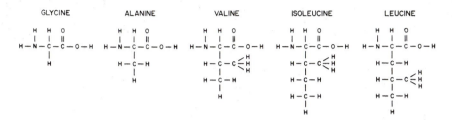

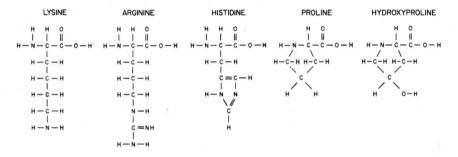

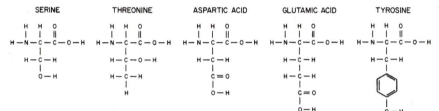

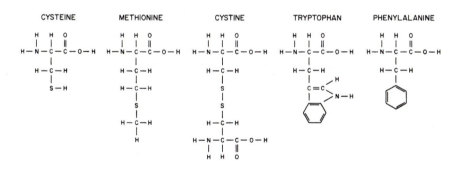

Figure 4.1 Amino acids: structural relationships.

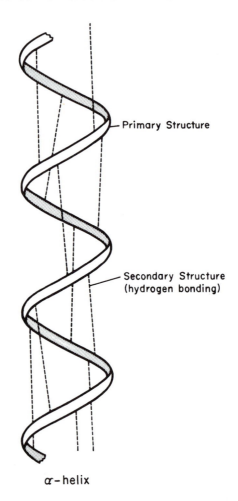

α-helix

Figure 4.2   Schematic model of a right-handed α-helix, as in the structure of proteins.

more peptide chains are tied together laterally by hydrogen bonding. Wherever hydrogen bonding occurs, a crystalline structure is observed.

Because proteins can be strung together in chains hundreds to thousands of units long, in different proportions, in assorted sequences, and with a great variety of branching and folding, an almost infinite number of different kinds of proteins is possible.

Cholesterol

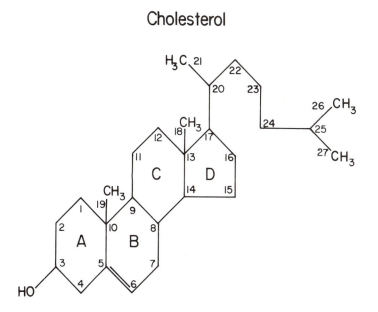

Figure 4.3 The structure of cholesterol.

LIPIDS

Associated with proteins in cellular membranes are the lipids or fats. Lipids are a rather heterogeneous class of compounds. They are classified as neutral lipids, phospholipids and sphingolipids, and glycolipids and terpenoid lipids, to which carotenoids and steroids belong. Simple lipids are esters of fatty acids and an alcohol. Cholesterol and fatty acids form esters that are found in living systems (Fig. 4.3). The widespread presence of lipids in microorganisms, plants, and animals has generated numerous experimental studies of their structural and metabolic roles in the cell.

Of the previously mentioned groups, phospholipids, and in particular lecithin, phosphatidyl choline, cephalin, and phosphatidyl ethanolamine, are the most abundant of the naturally occurring lipids in the cell. Their basic structure is illustrated in Figure 4.4. The most common phospholipids have chain lengths (represented by R_1 and R_2) of 16-20 carbons, with up to four carbon-carbon double bonds.

Phospholipids, fatty acids, and cerebrosides are all examples of polar lipids. They possess strongly polar or charged groups arranged in the molecules so that they may orient toward other polar molecules, for example,

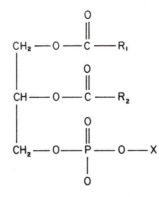

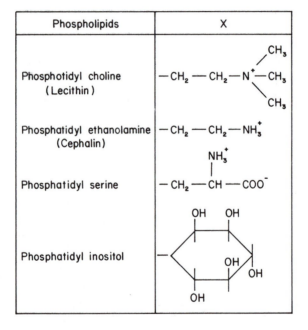

Phospholipids	X
Phosphotidyl choline (Lecithin)	$-CH_2-CH_2-N^+\begin{matrix}CH_3\\CH_3\\CH_3\end{matrix}$
Phosphatidyl ethanolamine (Cephalin)	$-CH_2-CH_2-NH_3^+$
Phosphatidyl serine	$-CH_2-\underset{\underset{NH_3^+}{\mid}}{CH}-COO^-$
Phosphatidyl inositol	

Figure 4.4 Basic structure of common phospholipids. R₁ and R₂ are akyl radicals with chain lengths of 16 to 20 carbons.

water. Their nonpolar portions are oriented away from the polar environment (Fig. 4.5). These lipids have melting points of 200-300°C, while lipids that have no polar groups in the molecule melt at much lower temperatures, around 70°C.

Most phospholipids disperse molecularly in water to only a small extent, and if larger quantities of phospholipids are introduced they aggregate to form micelles or liquid crystals. Whether one forms liquid crystals or micelles

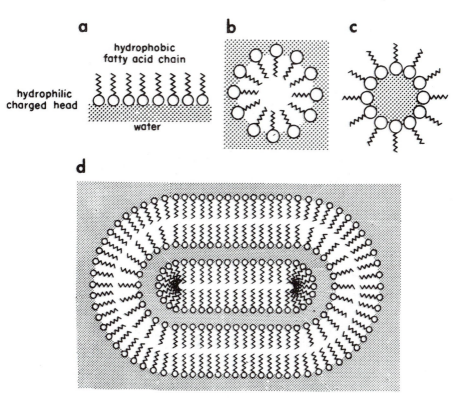

Figure 4.5 *(a)* A monolayer of phospholipid molecules on water. The phospholipids are symbolized by a circle representing the charged hydrophilic end, and the zigzag line represents the hydrophobic fatty acid chain. *(b)* If the liquid is polar, like water, the charged phosphates face outward. *(c)* If it is nonpolar, like benzene, they face inward. It can also exist as a combination of *(b)* and *(c)* as indicated in *(d)*. Refer to Figure 4.6.

depends primarily on the composition and temperature of the system. For example, lecithin, when mixed with water, undergoes a temperature-dependent phase transition from a turbid to a clear solution that exhibits the properties of a liquid crystal. When lecithin in water is observed under a microscope, it is seen to form myelin figures similar to those observed in living cells (Fig. 4.6). This observation will be discussed at greater length in the consideration of how molecules self-assemble into macromolecular structures, such as membranes, in chapter 5.

Lipid formations, particularly those of phospholipids, are of considerable physiological importance. Some of the most fundamental activities of life are inextricably entwined with their properties and structure. The cell

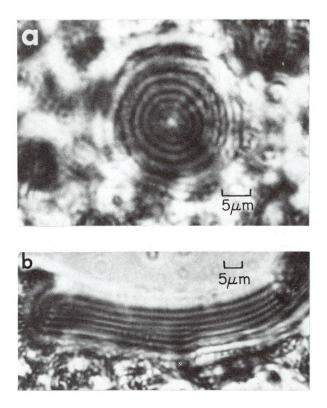

Figure 4.6 Lecithin (dipamityl lecithin) in water. Note lamellae. Compare to Figures 4.5*d* and 6.5 myelin.

membrane that regulates the inflow and release of molecules, the nerve cell membrane that controls the transmission of nerve impulses, cellular organelles such as the mitochondria that power the cell, and plant chloroplasts that trap radiant energy are only a few examples of vital living systems in which lipids function.

NUCLEIC ACIDS

Nucleic acids are essential in all living organisms to affect the processes of reproduction, growth, and differentiation. The nucleic acids are DNA (deoxyribonucleic acid) and RNA (ribonucleic acid). Both DNA and RNA are long chains of alternating sugar and phosphate groups. In DNA the sugar is deoxyribose; that is, the carbon at the 2′ position carries simply hydrogen, and in RNA the sugar is ribose, because the 2′ carbon carries a hydroxyl group. Each purine or pyrimidine, with its sugar and phosphate, is referred to as a nucleotide.

The nucleic acids are very large structures composed of aggregates of these nucleotides. The two purines, adenine and guanine, are present in both DNA and RNA. The pyrimidines commonly found in RNA are uracil and cytosine; in DNA, thymine and cytosine. Nucleic acids are pictured structurally as a double helix, and the DNA helical structure is illustrated in Figure 4.7. A comparison of this structure to that of the α-helix structure (Fig. 4.2) is informative. History of the research that led to the establishment of the α-helix structure for proteins and the double helix for DNA and to an understanding of the genetic molecules is summarized by Watson (1965), Olby (1974), and Judson (1979).

An almost endless variety of different nucleic acids is possible through variation of the nucleotide sequence, and specific differences among them are believed to be of the highest importance, for the nucleic acids are the main constituents of the genes, the bearers of hereditary information.

Transfer RNA is a small nucleic acid that plays a role in the cell by deciphering the genetic code. To understand these highly specific nucleic acid-protein or nucleic acid-nucleic acid interactions the structure of tRNA becomes important. The interaction of light and tRNA has given us a better physical picture of this molecule, but it has also revealed tRNA's light sensitivity (Favre and Thomas, 1981). In the bacterium *Escherichia coli,* growth can be delayed by exposing tRNA to near-ultraviolet light. We do not know as yet how the cells resume growth, but the repair mechanism of tRNA is of great interest.

We know that ultraviolet radiation from 200 nm to 300 nm is absorbed by nucleic acids and by proteins and that it brings about changes in these molecules. It is interesting that the ultraviolet wavelength at 260 nm corresponds to the absorption spectrum peak of DNA and that it increases the frequency rate of mutations.

Kelner (1949), studying bacteria, observed that ultraviolet-"killed" bacteria could be reactivated by visible radiation. Action spectra for photoreactivation show generally two peaks, one in the range of 300 nm to 380 nm and another from 430 nm to near 500 nm, depending on the organism. The process was found to be an enzymatic photocatalysis of pyrimidine dimers. Photoreactivation by visible light of ultraviolet damage continues to be

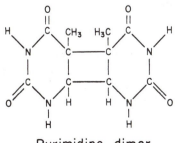

Pyrimidine dimer

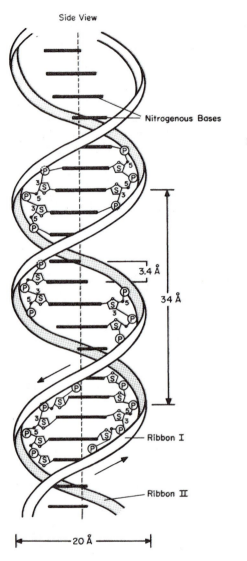

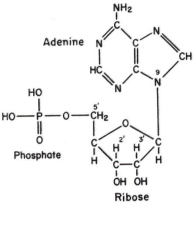

Nucleotide
(b)

DNA Molecular Structure
(a)

Figure 4.7 *(a)* The structure of the DNA molecule, that of double helix. This figure is a representation of the Watson-Crick model of the molecule (hydrated "B" form). Refer to Etkin, 1973, p. 653, and Watson, 1965. *(b)* A nucleotide structure, composed of a nitrogenous base adenine, a sugar ribose (or deoxyribose), and a phosphate.

effective and widespread in nature, for it has now been observed in algae, fungi, plants, and animals.

CARBOHYDRATES

A major source of our organic food is carbohydrates, many of which are derived from photosynthesis. In the process of photosynthesis carbohydrates are synthesized from carbon dioxide and water. More about this process is included in the discussion of photosynthesis in Chapter 8.

As food, carbohydrates provide an important source of energy, both as a means for storing energy and as structural molecules. Carbohydrates are classified as monosaccharides, disaccharides, and polysaccharides in accordance with the number of carbon atoms in their molecules. The monosaccharide glucose ($C_6H_{12}O_6$) is important in the process of cellular respiration. When two glucose molecules are linked they form the disaccharide sucrose ($C_{12}H_{22}O_{11}$). Polysaccharides are condensations of many monosaccharide molecules. For example, the monosaccharides in glucose can form long polymer chains such as starch, glycogen, cellulose, and chitin. It is interesting that the pentose sugars, ribose and deoxyribose, are part of the nucleic acids' molecular structure (Fig. 4.7*b*).

ADP-ATP SYSTEM

Another striking molecular feature all living organisms have in common is the presence of an adenosine diphosphate (ADP) and adenosine triphosphate (ATP) system as the energy storing mechanism (Fig. 4.8). This feature involves the synthesis of ATP from ADP and inorganic phosphate when surplus energy is being stored and the breakdown of ATP to ADP and phosphate when energy is required. Their particular molecular structures are well suited to give high energies of hydrolysis and in doing so supply energy for biochemical processes. The chemical energy of ATP is the key energy source for cellular function. It is used to perform all chemical and mechanical work of the cell, for example, biosynthesis, nerve conduction, muscular contraction, and locomotion.

The universal occurrence of the ADP-ATP system strongly suggests that it is the archetypical energy mechanism, a process that has been passed down from the earliest forms of life.

OPTICAL ACTIVITY

The structure and function of molecules of biological importance are related to their optical activity. For example, the optical activity of the amino acids

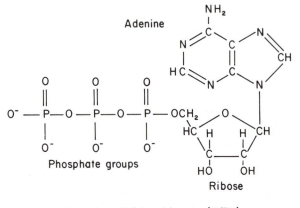

Adenosine Triphosphate (ATP)

Figure 4.8   Structure of adenosine triphosphate, ATP.

in polypeptide chains and ribose in nucleic acids is related to their ability to rotate the plane of polarized light in a certain direction. When the plane is rotated in a clockwise direction, the compound is said to be dextro-rotatory (D). If the plane of polarized light is turned in a counter-clockwise direction, the compound is said to be levo-rotatory (L). The two structures L and D optical isomers are mirror images; for example, L and D forms of the amino acid glycine.

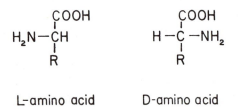

L-amino acid   D-amino acid

One would expect that the molecules from which life is believed to have arisen would contain nearly equal numbers of levo and dextro rotators called a racemic mixture. In fact, only L optical isomers occur in the amino acids in polypeptides and hence in the proteins of living organisms, while all sugars are found only in the D form. This finding has intrigued biochemists since Pasteur's early experiments with crystals of tartaric acid and sodium ammonium tartarate isomers. These experiments led Pasteur (1878) to an important observation—that asymmetric molecules are always the product of living processes. This fact appeared to Pasteur to be one of the fundamental differences between the chemistry of life and that of inanimate matter. He postulated that this peculiar asymmetry might be the manifestation of asymmetric forces in the environment acting on the evolving cells at the time of molecular synthesis.

The optical activity of organic compounds on the primeval earth could have been accelerated by an asymmetric catalyst. Experimental studies on the heat polymerization of simple organic molecules, such as polypeptides and ribose, have found that either the pure L or the pure D optical form is more stable than the racemic mixture. Most likely the L optical isomer was a chance happening in the first living organisms that evolved and was passed on through succeeding generations. Wald (1957, 1964a) expressed this as natural selection on the molecular level.

Looking back, we are struck by the fact that from relatively few atoms and simple molecules the largest, most complex molecular structures are built. No matter how complicated, no matter how diverse the chemistry of life became, these same molecules are found, from unicellular to multicellular organisms, and all the way to man himself.

5 *From Molecules to Cells*

The physiologist finds life to be dependent for its
manifestations on particular molecular arrangements.
Thomas H. Huxley, 1866.

Shortly after Darwin's publication of *The Origin of Species* (1859), Louis Pasteur (1860) showed by an ingenious set of experiments that life originated from existing life on Earth. The results dispelled for a time the belief that life arose as a spontaneous event. However, Pasteur did not rule out the spontaneous generation of life, for in 1878 he wrote: "I have been looking for it *(spontaneous generation)* for 20 years but I have not found it yet, although I do not think it is an impossibility." In this same context Darwin wrote in a letter to Joseph Hooker in 1871:

It is often said that all the conditions for the first production of living organisms are now present, which could ever have been present. But if (and—oh, what a big if) we could conceive in some warm little pond, with all sorts of ammonia and phosphoric salts, *light,* heat, electricity, etc., present, that a protein compound was chemically formed ready to undergo still more complex changes . . . . (Darwin, 1892, p. 220)

In their own ways Darwin and Pasteur dwelled on the possibility that life could have arisen as a spontaneous chance happening when the physical and chemical environmental conditions were right.

A new impetus for seeking the origins of life began to emerge in the late 1930s and was stimulated by the work of Oparin (1938) in *The Origin of Life,* which, once translated from Russian to English, found an eager audience and reawakened the thinking of scientists to the question of how life may have originated.

After the formation of the earth's crust, organic chemical formation began. The long span of time, which included the Archeozoic and Proterozoic geological eras, was the period of chemical evolution when relatively simple molecules were transformed into complex organic molecules (Fig. 3.1). Oparin (1938, 1968) envisioned that life arose from those preexisting compounds. About the same time Haldane (1928, 1954, 1966) in England suggested that organic compounds must have accumulated until primitive oceans reached the consistency of "hot diluted soup" before the origin of life. He based this suggestion on the assumption that the primitive atmosphere contained CO_2, NH_3, and water vapor, but no oxygen, and claimed that such a gaseous mixture exposed to ultraviolet radiation would give rise to a vast variety of hydrocarbons and organic molecules.

Assuming that the primordial atmosphere was the following gaseous mixture:

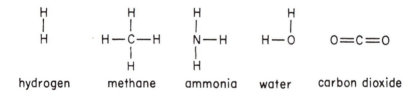

hydrogen   methane   ammonia   water   carbon dioxide

then some of the molecules that can be derived from them are:

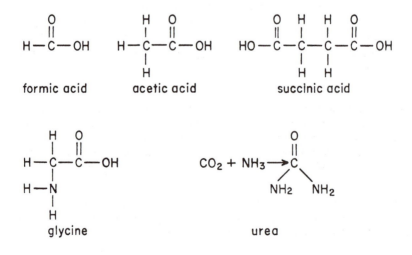

formic acid   acetic acid   succinic acid

glycine   urea

It is easy to see how from these molecules organic compounds could be synthesized without the agency of any living cells. The writings of Haldane (1928, 1966), Bernal (1951, 1967), Oparin (1938, 1968) and Urey (1952) pioneered modern thinking about how life in our universe originated.

At the suggestion of Urey, Stanley L. Miller (1957) performed a simple experiment. He filled the upper chamber of a flask with a mixture of methane, ammonia, and hydrogen gases and subjected the mixture to electrical sparking for a week. The upper chamber represented the ancient atmosphere of earth; lower chamber of the flask was filled with water to simulate the oceans. This process resulted in the formation of relatively high concentrations of amino acids, the basic components of all proteins. Variations of this experiment have been done many times using ultraviolet light, ionizing radiation, and high temperature, all resulting in the formation of a variety of essential amino acids. These experiments indicate that, given the right conditions, gases in the prebiological atmosphere could form the precursor molecules of life.

Calvin (1969, 1975), in looking for ways to synthesize complex organic molecules from simple molecules in a prebiological environment, irradiated carbon dioxide and hydrogen with a high energy ionizing radiation in the cyclotron. Formaldehyde, formic acid, acetic acid, and other reduced carbon compounds were obtained, which could then be used for further synthesis of a variety of complex organic compounds. Ponnamperuna (1972) and Oro (1980) showed how precursors of nucleic acids could have also formed under these primitive conditions.

In the prebiological environment there was considerable photochemistry resulting in the synthesis of many undetermined amino acids, protein-like molecules, lipids, and phospholipids and eventually leading to the generation of macromolecules, such as polypeptides, nucleic acids, and polysaccharides. All of these molecules were probably present in the primeval ocean.

Given the wealth of biological precursor molecules, scientists turned to the question of how these molecules could have organized themselves out of this original organic "soup" into a living cell. It is estimated that progress from such beginnings to a living cell capable of growth and replication took about 3.5 billion years. A possible hypothesis for the evolution of life, with time and environmental changes as they may have existed, was conceived by Gaffron (1965) as shown in Table 5.1.

How did the molecules in the prebiological environment self-organize to form an assembly of macromolecules? Let us briefly examine the various bits of experimental data suggestive of a physical-chemical basis for the origin of life. Bernal (1951) suggested that clay could provide a sufficient surface for organic molecules to be adsorbed and that in the presence of a catalyst the synthesis of macromolecules could have rapidly taken place. Clay may have

Table 5.1 Evolutionary Eras in the Development of Life

Era	Environment	Energy Source	Outcome
I	Anaerobic CH_4, NH_3, H_2	ultraviolet light high temperature	acetate, glycine, uracil, adenine "organic soup"
II	Loss of Hydrogen anaerobic; traces of O_2	ultraviolet light high temperature visible light	polyphosphates, peptides, porphyrins oxidation-reductions
III	Loss of Ultraviolet anaerobic; traces of O_2, CO_2	visible light	surface catalysis, photochemistry synthetic reproduction cycles
IV	Loss of Free Food anaerobic; CO_2, traces of O_2	photoreduction fermentation	multiplying metabolic cells
V	Loss of Anaerobiosis aerobic; anaerobic pockets	photosynthesis respiration	autotrophic plants

Source: After Gaffron, 1965, p. 450.

also played a role in the very early polymerization reactions (Oro, 1980). Vinyl monomers, either adsorbed on the external surfaces or between the lamellae of clay, do indeed polymerize. The hypothesis of crystalline-organic particles bound by adsorption of organic molecules onto surfaces of mineral clay is supported by recent findings in which amino acids were isolated from meteorites.

Molecules can be brought together by being built into a crystal, in which case the lattice forces will hold the molecules together. In the process of crystal growth, the progressive accretion of molecules on the crystal surface could have served as a template for a self-replicating system. Cairns-Smith (1971) developed a model in which the first functional macromolecules, proteins, replicated on templates. These templates may have been microscopic crystals of various kinds of clay, containing trace metallic ions that could have acted as catalysts. Cairns-Smith (1982) further points out that the adsorption of amino acids or protein chains to the surface of a particular template will be specific, depending on the configuration of the organic molecules. That is, aggregates of clay particles and protein chains could have specific properties and could react to changes in the environment.

Another approach is that organic molecules separated at different phase boundary layers in the sediment strata during certain temperature stages in the cooling of the earth's surface. Evidence is found among the layers where organic materials, hydrocarbons, and phospholipids are found. The phospholipids could have served as a boundary membrane for the synthesis of more complex organic molecules. The hydrocarbons were not only present in clay but were also on oceanic surfaces, and when subjected to ultraviolet radiation could have led to a phospholipid-water bilayer (Fig. 4.5). These bilayer membranes might in turn have formed microspheres. Therefore, radiation and temperature variations were key factors in the synthesis, polymerization, and aggregation of organic molecules into macromolecular protocells. So we see that primordial cells could very well have spontaneously arisen in the early environment.

Attempts to create structures that resembled cells from a variety of organic and inorganic molecules go back to the experiments of Leduc (1911). The possibility of building high molecular weight compounds from dilute colloidal suspensions was of interest to Bungenberg de Jong (1936). Oparin (1938) applied these studies in his search to find a dynamic colloidal system as a model for the origin of life. One such model system was originally made by mixing gelatin with gum arabic in water. At 42°C a clear solution was obtained. Other solutions were made with gelatin and lecithin and a variety of other substances. The important observation Oparin made was that at a critical pH, microspheres came out of solution; he named these *coacervates*. These were colloidal microspheres with osmotic properties similar to those of living cells. Oparin and his associates have continued to experiment with this model system demonstrating that they can perform a variety of biochemical reactions.

A series of experiments based on the fact that amino acids were present in the prebiological environment was begun by Fox in the early 1960s. He concentrated on the polymerization of amino acids. Initially, the method consisted of heating a mixture of amino acids to temperatures of 160–200°C for several hours under anhydrous conditions in an atmosphere of nitrogen. The mixture contained aspartic acid, glutamic acid, and lysine (Fox, 1965*a*, 1965*b*; Fox et al., 1963). Using thermal polycondensation they were able to copolymerize amino acids, which were named *proteinoids*. When dry proteinoid was treated with hot water, microspheres separated out of the cooling clear solution (Fig. 5.1). The proteinoid microspheres varied in size from a few microns to 100 μm in diameter, were relatively stable, and, depending on their preparation, exhibited living-cell-like behavior. They possessed a cell membrane, budded, coalesced, increased in size, and divided.

Fox and his collaborators have demonstrated that proteinoids are protocells that are able to catalyze a considerable number of biochemical reactions as do living cells. For example, given the right conditions, the proteinoid

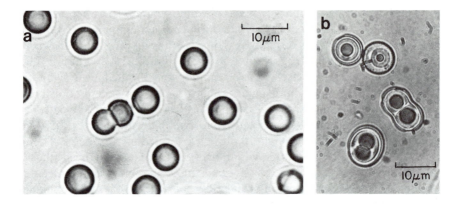

Figure 5.1   Protenoid microspheres, prepared by the polymerization of amino acids. (Courtesy of S. Fox, Institute of Molecular Evolution, University of Miami, Miami, Florida.)

protocell can synthesize ATP, make peptides, and catalyze the reaction of nucleotides to form polynucleotides, which are precursor molecules on the pathway to making DNA. The sequence of events that may have occurred is conceptualized by Fox (1980) in Figure 5.2.

LIQUID CRYSTALLINE SYSTEMS

> . . . living systems actually are liquid crystals or it would
> be more correct to say, the *paracrystalline* state
> undoubtedly exists in living cells.
> Joseph Needham, 1950.

A model that applies to the question of how molecules became self-ordered and assembled into macromolecular structures is that of liquid crystalline systems (Brown and Wolken, 1979). Such models have not yet been explored in depth concerning the problem of biogenesis.

Let us describe briefly what liquid crystals are and the kind of structural properties they possess. Friedrich Reinitzer (1888), an Austrian botanist, discovered the liquid crystalline state. He prepared the cholesteryl ester, cholesteryl benzoate, and observed that it has two melting points with different properties. At 145.5°C, the solid cholesteryl benzoate structure collapsed to form a turbid liquid (now known to be a liquid crystal), and when further heated to 178.5°C, it became transparent. Lehmann (1904)

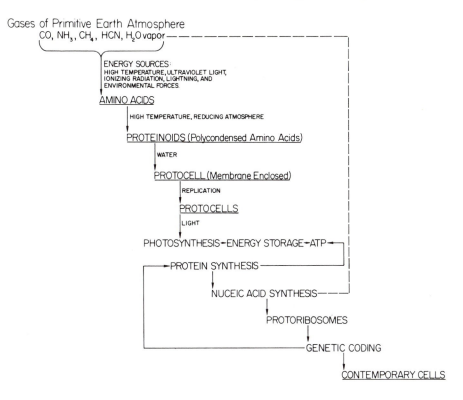

Figure 5.2   From gases of the primitive atmosphere to protocells and to contemporary cells. (Hypothetical scheme adapted from Fox, 1980.)

made a systematic study of organic compounds and found that many of them exhibited properties similar to cholesteryl benzoate, in that the compounds behaved both as a liquid in their mechanical properties and as a crystalline solid in their optical properties. Lehmann (1904, 1922) originated the term *liquid crystal* and recognized that these properties exhibited by liquid crystals can in fact be found in living cells. Then Rinne (1933) and Bernal (1933, 1951) pointed out that naturally occurring liquid crystals are intimately connected with life processes.

There are two major types of liquid crystals: the thermotropic and the lyotropic. Thermotropic liquid crystals are formed by heating, and lyotropic liquid crystals are formed by mixing together two or more compounds. The thermotropic liquid crystals are themselves divided into two groups described as *nematic* and *smectic* (Fig. 5.3).

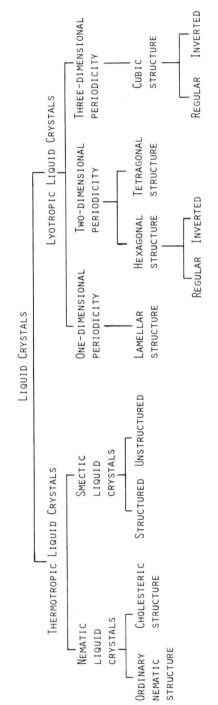

Figure 5.3 Classification of liquid crystals. *(Based on classification from Brown, 1977, and modified by Brown and Wolken, 1979, p. 13.)*

Nematic liquid crystals are structurally different from isotropic liquid crystals in the spontaneous orientation of the molecules along their long axes. In the nematic form the molecules maintain a parallel or nearly parallel arrangement to each other. They are mobile in three directions and can rotate about one axis. The smectic structure is stratified with the molecules arranged in layers. Their long axes lie parallel to each other in the layers, approximately perpendicular to the plane of the layers. The molecules can move in two directions in the plane, and they can rotate about one axis. Within the layers the molecules can be arranged either in rows or can be randomly distributed (Fig. 5.4*a, b*).

Included with the nematic liquid crystals is a subclass that is considered *cholesteric-nematic,* but is referred to as *cholesteric.* Many of these compounds are derivatives of cholesterol. In the cholesteric liquid crystals, the molecules pack in layers. Although most of the molecules in this state are essentially flat, side chains project upward from the plane of each molecule with some hydrogen atoms extending below. Thus, the direction of the long axis of the molecules in a chosen layer is slightly displaced from the direction of the axis in adjacent layers and produces a helical structure (Fig. 5.4*c*).

A nematic liquid crystal can be converted to a cholesteric liquid crystal by adding a molecule that has an asymmetric atom. Cholesteric liquid crystals (cholesteryl esters) are miscible with nematic liquid crystals forming a helical structure. We see this same pattern of cholesteric helical structure in

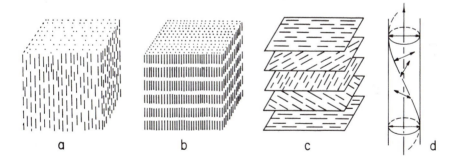

a   b   c   d

Figure 5.4 Structure of the molecular arrangement in three main types of liquid crystals:
    a) Nematic, the elongated molecules are randomly distributed.
    b) Smectic, molecules are ordered in layers (lamellae) of equal thickness.
    c) Cholesteric, the molecules rotate regularly from plane to plane and
    d) helix of a cholesteric rod.
(From Wolken, 1984, p. 144, modified from Brown and Wolken, 1979, pp. 6 and 24.)

important life-associated molecules already mentioned: in polysaccharides, polynucleotides and polypeptides, DNA, and RNA. They all assemble in cholesteric helical structures.

The lyotropic liquid crystals differ from thermotropic liquid crystals in that they are obtained by mixing one compound with another, one of which is a solvent. They are strongly birefringent. Most detergents, soaps, and surfactants dispersed in water will form lyotropic liquid crystals (Fig. 5.3). These compounds are *amphiphiles,* meaning they possess in their molecular structure an ionic group that is water soluble and an organic part that is insoluble in water.

Starting with the crystalline amphiphile and water one can generate a full range of structures from a true solid to a true liquid. Removing the water can reverse the order of formation. By using combinations of specific amphiphiles one can generate more complex structures, either lamellar, cubic and hexagonal. These structural transformations are illustrated as follows:

$$\text{Solid} \overset{+H_2O}{\underset{-H_2O}{\rightleftharpoons}} \begin{matrix}\text{Liquid} \\ \text{crystal} \\ \left\{\begin{matrix}\text{lamellar} \\ \text{structure}\end{matrix}\right\}\end{matrix} \overset{+H_2O}{\underset{-H_2O}{\rightleftharpoons}} \begin{matrix}\text{Liquid} \\ \text{crystal} \\ \left\{\begin{matrix}\text{cubic} \\ \text{structure}\end{matrix}\right\}\end{matrix} \overset{+H_2O}{\underset{-H_2O}{\rightleftharpoons}} \begin{matrix}\text{Liquid} \\ \text{crystal} \\ \left\{\begin{matrix}\text{hexagonal} \\ \text{structure}\end{matrix}\right\}\end{matrix} \overset{+H_2O}{\underset{-H_2O}{\rightleftharpoons}} \text{Micellar} \overset{+H_2O}{\underset{-H_2O}{\rightleftharpoons}} \text{Solution}$$

In the cubic structure the molecules pack in a spherical pattern, and the spheres then pack in a cubic design. In the hexagonal structure the molecules pack in a cylindrical pattern, and the cylinders or rods pack hexagonally.

Lyotropic liquid crystals are biologically important, for they contain a system of two or more compounds (e.g., lipid-water; lecithin-cholesterol-bile salts-water; lipid-water-protein) in which water is an integral part. Cellular membranes are structural bilayers of a lipid-protein-water system; they are in fact liquid crystalline structures.

The general classification of liquid crystals is indicated in Figure 5.3 and these liquid crystalline structures are schematically illustrated in Figure 5.4. In the nematic state the long axes of the molecules lie essentially parallel (Fig. 5.4a), while in the smectic A structure (Fig. 5.4b), the molecules show two-dimensional order. Within a layer the molecules are randomly distributed, while between layers the molecular arrangement is equally spaced. The molecules in smectic C liquid crystals are packed in equidistant layers as in the smectic A liquid crystals, but the molecules in a given layer are tilted in relationship to the plane of the layer. The tilt angle is sensitive to temperature and the molecular geometry of the molecules. In the cholesteric structure the molecules are arranged in each layer like those in the nematic structure, but a necessary twist is superimposed on the layers resulting in a helical structure (Fig. 5.4c) or a helical rod (Fig. 5.4d).

The single most important characteristic of liquid crystals is that they possess, simultaneously, both mobility and structural order. Like living cells, liquid crystals are sensitive to a variety of external stimuli: light, mechanical pressure, temperature, electric and magnetic fields, and changes in the chemical environment.

The structural molecules of living cells are lipids, proteins, and water. Low molecular weight proteins with water can generate liquid crystalline structures, especially if the protein possesses a high degree of polarity. As we have stated, proteins are composed of amino acids structured together via peptide linkages that form polypeptides. Polypeptides have been an extensively studied class of liquid crystals (Dupre and Samulski, 1979). Robinson (1956, 1958, 1966) carried out a series of classical studies of the synthetic polypeptide, poly-γ-benzyl-L-glutamate (PBLG). PBLG is a homopolymer of a single amino acid with the basic repeat unit shown below:

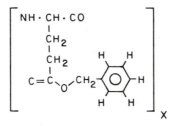

PBLG in chloroform forms a large number of spherulites upon evaporation. The spherulites coalesce and yield a birefringent material that spontaneously orients itself at the air interface. The polypeptide-rich phase is birefringent and separates from the more dilute medium in the form of spherical liquid droplets or spherulites. If the two-phase solution is cooled or the polypeptide concentration further increases, the droplets grow in size and coalesce, forming a birefringent fluid, a lyotropic liquid crystal. Spherulites (Fig. 5.5), like proteinoid microspheres, are enclosed by a membrane and divide by passing through an oblate spheroid phase. This self-orientation exhibited by spherulites provides a model then for how molecules form oriented structures that are of biological importance.

The aggregation of chain molecules is important in the chemistry of liquid crystalline structures, in the formation of micelles, and as a possible mecha-

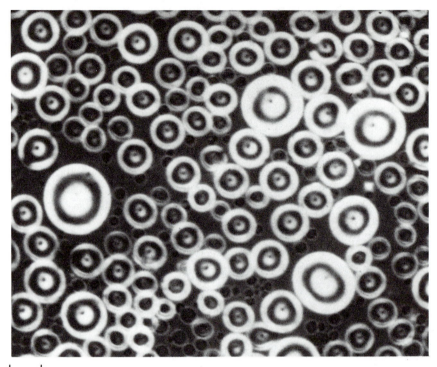

10 μm

Figure 5.5   Spherulites formed from poly-γ-benzl-L-glutamate in chloroform, near the transition temperatures between nematic liquid crystals and the isotropic liquid. The spherulites are in a dynamic state of transition. The small spherulites coalesce with larger ones and then divide. *(From Brown and Wolken, 1979, p. 82.)*

nism by which the structural formation of cellular membranes originally took place. The cell membrane separates the internal environment from the external environment and provides a large surface area for bringing molecules together for interaction and energy transfer. The critical question is: How did the hydrocarbons in the environment self-order to form a cell membrane capable of encapsulating other molecules from the environment for metabolic processes?

The major molecules in the lipid bilayers are phospholipids, and the molecular packing of these lipids dictates the skeletal structure of the membrane. Phospholipids are polar lipids (Fig. 4.4).

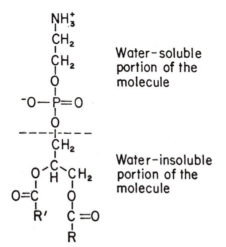

NH_3^+
|
CH_2
|
CH_2 Water-soluble
| portion of the
O   molecule
|
$^-O—P=O$
|
O
- - - -|- - - -
|
CH_2 Water-insoluble
| portion of the
$O^{^-}C^{^-}CH_2$ molecule
O=C |
| O
R' |
$C=O$
|
R

In water the charged phosphates face outward, and if the medium is nonpolar, they face inward (Fig. 4.5). In other words, the molecules orient toward water or other polar molecules, and their nonpolar groups orient away from the polar environment. Most phospholipids disperse molecularly in water to only a small extent, and if large quantities are introduced into the aqueous medium, micelle aggregates form. Phospholipids swell in water and form spherical bodies composed of concentric layers (lamellae) with water trapped between them. If the spheres are surrounded by a single phospholipid bilayer, they are referred to as *liposomes*. Lecithin dispersed in water will form concentric bilayered lamellae, which are observed as myelin structures in cells. These lipid molecules will self-assemble into ordered, replicating structures resembling cellular membranes, and need no further regulation or addition to align into an ordered structure. For example, lecithin in physiological saline will self-align into a precisely ordered replicating structural membrane, as observed in Figure 5.6. All of these phospholipid mono- and bilayer membrane structures are liquid crystalline systems (Chapman, 1973, 1979).

Phospholipids are extremely temperature sensitive. Upon heating they undergo an endothermic transition at a temperature well below their melting point, which would be at about 200°C. At this transition temperature a change of state occurs from the crystalline or gel to the liquid crystalline state. This change is associated with increased conformational freedom for the lipid fatty acid chains. The transition temperature rises with increasing length of the fatty acid chain and lowers with increasing unsaturation in the chain (Lee, 1975). Below the transition temperature, in the gel (crystalline) phase, the phospholipids adopt a bilayer structure in which the fatty acid chains

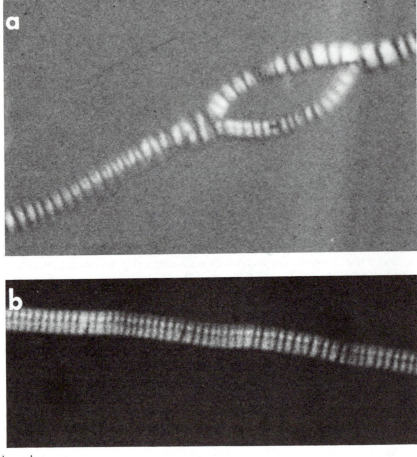

5 μm

Figure 5.6 Membrane formed from lecithin in physiological saline, using polarized light, a quarter-wavelength filter, and photograph taken by phase contrast microscopy. *(Refer to Wolken, 1984, p. 155.)*

are packed in ordered hexagonal arrays where motion of the fatty acid chains is highly anisotropic and restricted. At the transition temperature, there is a 50% increase in the surface area occupied by the lipid, and appreciable motion becomes possible about the C-C bonds of the fatty acid. The motion about the C-C bonds in lipid bilayers has been studied by ^{13}C NMR, and

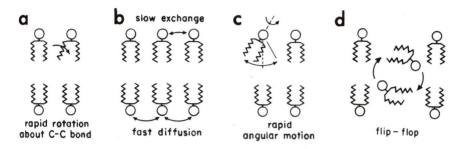

Figure 5.7 Schematic showing mobility of the phospholipid hydrocarbon chain in the cell membrane.
(a) Rapid rotation about C-C bond in the hydrocarbon.
(b) Lateral diffusion in the plane of the membrane.
(c) Angular motion of the phospholipid molecule.
(d) Flip-flop of the phospholipid molecules across the bilayer.
(From Brown and Wolken, 1979, p. 158.)

shows resonances for many of the fatty acid chain carbons. It was found that there is a distinct motional gradient within the phospholipid molecule in the liquid crystalline phase. The motion abut the C-C bonds with the fatty acid ^{13}C NMR data provided evidence suggestive of an axial rotation of the whole lipid molecule in the plane of the bilayer. The degrees of freedom possible for the phospholipid chains are schematized in Figure 5.7.

One can say that a membrane containing phospholipids with little unsaturation is less fluid than one with much unsaturation. The control of fluidity of the components of cell membranes may be related to the diffusional characteristics of molecules and ions passing in and out of the membrane. The state of the phospholipid in a membrane, in a gel, or in a liquid crystalline state can be expected to have a marked effect on the function of the membrane. Thus, small molecules will be able to move relatively easily through a membrane in which the phospholipids are in a liquid crystalline state.

... the oriented molecules in liquid crystals furnish an ideal medium for catalytic action, particularly of the complex type needed to account for growth and reproduction ... a liquid crystal has the possibility of its own structure, singular lines, rods and cones, etc. Such structures belong to the liquid crystal as a unit and not to its molecules, which may be replaced by others without destroying them, and they persist in spite of the complete fluidity of the substance. They are just the properties to be required for a degree of organization between that of the continuous substance, liquid or crystalline solid, and even the simplest living cell. (Bernal, 1933, p. 1082)

LIQUID CRYSTAL CELL MODEL

The properties and structures described for various liquid crystals suggest how a liquid crystalline system may have organized into a protocell. The characteristics of living cells are their stability, their response to environmental stimuli, and their ability to reproduce exact copies of themselves. Simple chemistry cannot account for such behavior.

The chemical and mechanical properties of liquid crystals are applicable in considering a cell model that has a solid core of nematic liquid crystalline material and is surrounded by a smectic liquid crystal (Fig. 5.8). The interior of the cell is a nematic liquid crystal with a number of important features that may be utilized. In the self-ordering nematic structure, the long axes of the molecules are essentially parallel (Figs. 5.6 and 5.7), and as they line up, they generate a layer of molecules that serves as a substrate on which chemical reactions can take place. Simple organic reactions such as isomerization and enzymatic oxidation-reduction reactions can occur on these liquid crystalline surfaces. Even small chemical changes result in changes of shape, and internal distribution of molecules will be accomplished at a much greater rate than would be possible by other states of matter. The surface energy in a droplet of nematic liquid crystal is not a constant but varies as a function of molecular alignment of the surface.

The smectic liquid crystalline phase has many dynamic properties just like the nematic phase. The cell membrane that has as its principal component a smectic liquid crystal would be able to transfer ions across it. If a smectic film contacts a region that lowers its surface energy, it will expand its

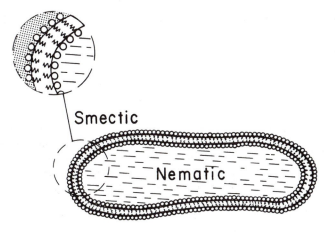

Smectic

Nematic

Figure 5.8 A schematic of a liquid crystal protocell. *(From Wolken, 1984, p. 145.)*

area of contact. Thus, a smectic film will engulf materials from the environment that tend to lower its surface energy.

The shape of liquid crystalline droplets that are uniformly aligned may exhibit a variety of forms, from an oblate to a prolate spheroid, depending on the properties of the liquid in which they are immersed. The liquid crystalline structure responds readily to energy changes, and remarkably, the model could have many other behavioral properties of a living cell (Fergason and Brown, 1968; Brown and Wolken, 1979; and Wolken, 1984).

REMARKS

We do not know at present how life originated on Earth. Current thinking about the origin of life supports the view that life arose as a spontaneous event when the molecules in the environment found the right conditions to self-order and replicate into stable structures.

The diurnal alterations of solar radiation and temperature on the earth's surface were important factors in creating special conditions for the synthesis of organic molecules associated with living organisms. The physical and chemical forces in the environment shaped these molecules into particular molecular geometries that self-aligned and replicated into macromolecular assemblies. Their stability was dictated by their chemical constituents.

Our experimental model systems have given us some clues and insights into how life could have originated on Earth. These models simulate many behavioral and biochemical processes associated with a living cell. However, much crucial information on the physical-chemical origins of life remains to be deciphered and much more experimentation is necessary. It is a long way from the chemical evolution of organic compounds and macromolecular structures to the origin of life and biological evolution.

6

The Cell

> It must be understood that no matter how minute an
> organism may be or how elementary it may appear at
> first glance it is nevertheless infinitely more complex
> than any simple solution of organic substances. It
> possesses a definite dynamically stable structural orga-
> nization which is founded upon a harmonious combina-
> tion of strictly coordinated chemical reactions.
> Oparin, *Origin of Life,* 1938.

THE LIVING CELL

All life shares a common ancestry, for all forms of life as we know them arose
from a cell, some 3.5 billion years ago. This cell gave rise to greater complexity by evolving metabolic pathways and structures to carry on life processes.
The living cell is a complex dynamic system enclosed within a membrane,
involved with the processes of energy, processes vital to its maintenance,
growth, and reproduction.

Robert Hooke (1665), using a simple microscope, visualized that the cell is
the unit of life and that its existence depends on a cell membrane. The origin
of the cell theory, though, is credited to Schleiden (1838) and Schwann
(1839), who stated that the cells are organisms arranged in accordance with
definite laws. Rudolf Virchow in 1858 recognized that each cell originates
from another cell, and this discovery marked the beginning of the science of
cellular systems. Claude Bernard in 1866 pointed out that the organic matter
in a cell is very specially organized and that this structured organization is
necessary for living cells.

Developments in optics have greatly improved our ability to observe
details in the cell. Microscopic methods for phase, polarization, interfer-

ence, and fluorescence have revealed considerable information about various types of cells. Electron microscopy, X-ray and optical diffraction, spectroscopy, computer imaging, and other electro-optical instruments have brought us ever closer to the molecular description of a cell.

Cells vary in shape, size and internal organization. Depending on their structure they are described as primitive or advanced cells. Primitive cells are all species of bacteria and cyanobacteria (blue-green algae). They do not contain a membrane-enclosed nucleus and their genetic molecules are dispersed throughout their cytoplasm. Lacking a nucleus, they are *prokaryotes* (Fig. 6.1). Cells that possess a membrane-enclosed nucleus are *eukaryotes*. These cells, in addition to a nucleus, evolved other specialized organelles such as mitochondria, endoplasmic reticulum with ribosomes, chromatophores, photoreceptors, and other structures (Fig. 6.2). These specializations enabled cells to adapt to their environment and maximize their energy usage for survival. Eukaryote cells are found in all more highly evolved plants and animals. These various types of cells aggregated for mutual benefit to produce a whole integrated multicellular plant or animal. We will now more closely examine the structure and components of these cells.

THE CELL MEMBRANE

The cell membrane is an important, integral component of the cell and it will be useful to describe its generalized structure before discussing other spe-

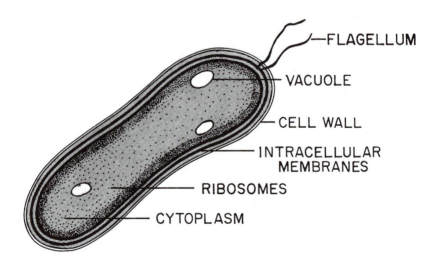

Figure 6.1 Schematized bacterial cell. Prokaryote.

cific cellular membranes. The cell membrane provides the means to separate the external environment from the internal environment of a cell. In order to allow for the differential diffusion of ions and the exchanges of gases, it has selective properties. The cell membrane was at one time envisaged as a passive barrier to diffusion and permeability, but it is now known to play an active role in chemical transport, energy transduction, and information transfer to and from the cell.

Cellular membranes depend on the physical-chemical properties of lipids and proteins. For most membranes, lipids (sterols and phospholipids) are found in concentrations greater than 30%. The lipids have the unique property of forming mono- or bimolecular layers when dispersed in water (Figs. 4.6, 5.6 and 5.7), because of the presence of hydrophilic (water soluble) groups at one end of the molecule and hydrophobic (lipid soluble) groups at the other end of the molecule. Cell membranes studied by electron microscopy and X-ray diffraction indicate that the cell membrane structure is a bilayer, 100 Å in thickness, each layer of the bilayer being about 50 Å thick (Fig. 6.3). Proteins and enzymes are molecularly associated with the lipid bilayers. The most simple cell membrane molecular structure is schematized in Figure 6.4a and is referred to as the *unit membrane*. However, no one molecular model applies to all cellular membranes, for cellular membranes are not static but dynamic structures and their molecules have mobility. Examples of other molecular models of the cell membrane and how the proteins may fit into the lipid bilayer are shown in Figure 6.4a-f.

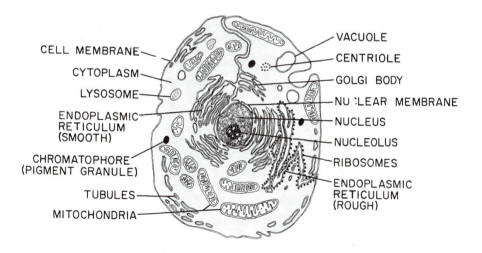

Figure 6.2 Schematized animal cell. Eukaryote. Refer to Figure 6.5.

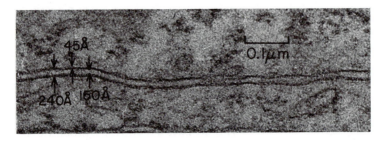

Figure 6.3   A cell membrane. Electron micrograph. Refer to Figures 5.6 and 6.5.

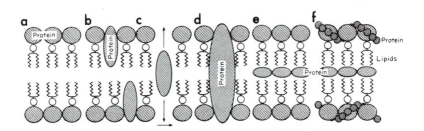

Figure 6.4   Cell membrane models. *(a)* The lipid bilayer with associated proteins. *(b)* The lipid bilayer in which proteins are not only on the surface, but another protein lies between the bilayer. *(c)* The protein with rotational reaction as well as lateral diffusion. *(d)* A membrane in which a protein extends above and beyond the lipid bilayer. *(e)* The lipid bilayer in which the protein not only is on the surface, but another protein lies between the lipid bilayer. *(f)* A membrane similar to *(e)*, but another protein is wrapped in an α-helix around the surface protein.

THE NUCLEUS

The nucleus is the most conspicuous body in eukaryotic cells. It is separated from the cell cytoplasm by a nuclear membrane (Fig. 6.5). In some cells the center of the nucleus is occupied by a large ovoid body, the nucleolus. Cells that undergo nuclear division have a set of chromosomes containing within their genes the hereditary molecules DNA and RNA. The chromosomes are most pronounced during nuclear division and undergo a process known as mitosis in which the chromosomes split and separate during cellular reproduction. Mitosis accelerated the process of evolution by incorporating chance mutations directly into the genetic future of a cell through random mutation of nuclear DNA.

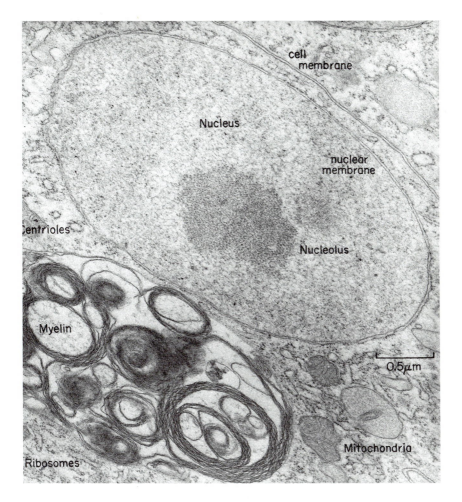

Figure 6.5 Section through a cell. Note cell membrane, nucleus, nuclear membrane, mitochondria, myelin, centrioles, and ribosomes in cytoplasm. *Daphnia pulex*. Compare to Figure 6.2.

MITOCHONDRIA

The mitochondria are the most numerous of the cytoplasmic organelles (Fig. 6.6). They are the site of chemical reactions and contain the cytochromes, the respiratory enzymes for the cell.

Structural studies by electron microscopy indicate that mitochondria possess a limiting or outer membrane and an inner membrane. A system of ridges protrudes from the inside surface of the inner membrane and these ridges have been designated as *cristae*. It has been suggested that the

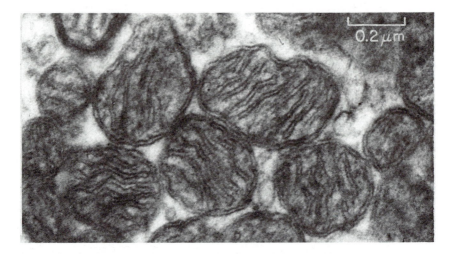

Figure 6.6   Mitochondria, in inner segment of retinal rod cell of frog. *(From Wolken, 1966b)*

oxidative enzymes of the mitochondria are built into these cristae. The inner membrane consists of an assembly of closely packed repeating units, represented by bilayers of lipid and protein. Phospholipids make up 40-60% of the dry weight of mitochondria. These membranes are the transducers in the process of oxidative phosphorylation. In this process, electrons move along an electron transfer chain, and are linked to the synthesis of adenosine triphosphate (ATP) whereby energy is stored. Mitochondria also have their own DNA and RNA, indicating that they are self-replicating structures along with the cell. The mitochondria structure is schematized in Figure 6.7.

ENDOPLASMIC RETICULUM AND GOLGI

The endoplasmic reticulum is an organization of membranes (Fig. 6.2). This complex cytoplasmic system of vesicles and tubules was first revealed by electron microscopy, which showed that its elements may be integrated in a continuous network. It contains the system for protein synthesis with which ribosomes for RNA synthesis are associated. Another cluster of membranes with numerous vesicles is called the *Golgi body* or apparatus (Fig. 6.2).

MYELIN

A variety of membranous structures are found in cells. One of these is described as myelin figures, which consist of concentric membranes that are

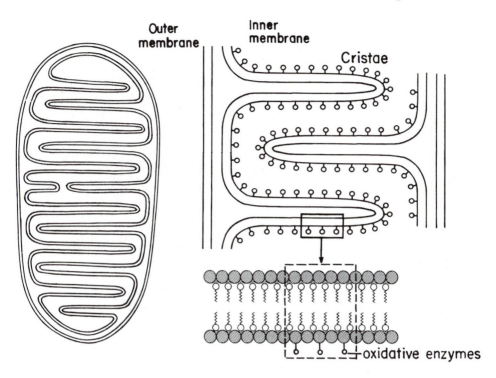

Figure 6.7 Mitochondrion. Structural model.

bilayers of phospholipids (Figs. 6.5 and 6.8). Myelin figures have been used as models for studies of the cell membrane structure. Nerves have a myelin sheath, a biomolecular layer of lipid and protein that surrounds the axon of the nerve cell.

CILIA AND FLAGELLA

The cilia and flagella propel the cell through its medium and if the cell is fixed in place, the cilia move things past it. In the tissue cells of higher animals, flagella and cilia have become adapted to functions associated with memory structures.

The flagellum consists of a number of elementary filaments (axonemata) embedded in a matrix and covered by a membrane (Fig. 6.9). There are eleven elementary filaments, of which nine (paired microtubules) are peripherally located, while the other two are sometimes found in the center of the flagellum. The pattern is known as the *9 + 2 array* and is associated with all

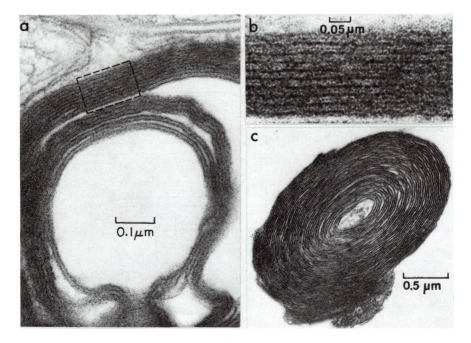

Figure 6.8 Myelin structures in cells *(a, b, c)* and that which surrounds a nerve axon. Electron micrographs.

motile cells. This arrangement is observed in a variety of plant and animal cells with flagella or cilia, from bacteria to the sperm tails of man (Fig. 6.10).

Among prokaryotes, the flagella and cilia are similar. They are small and single stranded, consisting of a protein called *flagellin*. Electron microscopy of bacterial flagella shows that they are constructed of globular subunits and that the subunits are arranged in helices of various kinds, depending on the type of bacterium studied. From X-ray diffraction and other studies, the bacterial flagellin protein shows an α-helix pattern (Fig. 6.10c). The structure of bacterial flagellin led Astbury (1933) to suggest that it could be regarded as a "monomolecular muscle."

Microtubules from any kind of eukaryotic flagella or cilia are composed of a protein called *tubulin*. At the base of every eukaryotic flagellum and cilium is a distinct microtubular structure called the *basal body*. The architecture of the basal body is identical to that of the *centriole,* a structure found at opposite poles of the eukaryotic cell nucleus. Centrioles are found in nearly all animal cells, in the cells of many eukaryotic algae, and in most higher plants. Centrioles come into particular prominence during mitosis.

The structural array of the basal body and the centriole is *9 + 0*; the

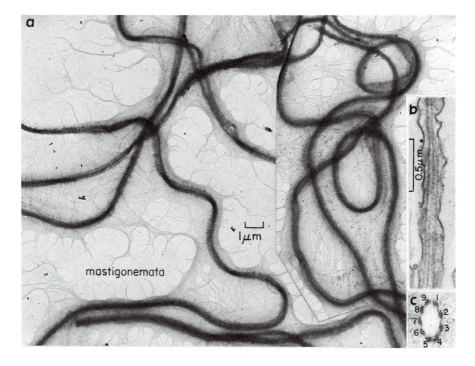

Figure 6.9 Flagella *(a)* isolated from *Euglena gracilis,* not sectioned, but stained with osmium tetroxide (OsO₄). Sections through flagellum *(a)*, longitudinal *(b)*, and cross section *(c)*.

central pair of microtubules is absent. In cells that possess mitotic centrioles, the centrioles left over from earlier divisions often grow projections that become flagella or cilia as the new cell differentiates. Thus, not only are the basal bodies and centrioles identical in structural pattern, but centrioles can also become basal bodies. Moreover, the mitotic spindle, the structure that lies between the centrioles during cell division, is an array of microtubules composed of the protein tubulin. It is interesting from an evolutionary as well as a structural standpoint to find microtubules so widely distributed in all types of cells.

There are other cellular inclusions that should be noted such as vacuoles, pigment bodies, granules, and crystals. Crystals have been identified in bacteria protozoa, fungi, plants, and animal cells. The crystals isolated from these cells are primarily protein, but lipids and pigments are sometimes associated with them. In animal cells, crystals are of considerable interest, for they are found in the sex glands, the heart, and in other cellular tissues of aged animals, but their function in these cells is not presently understood.

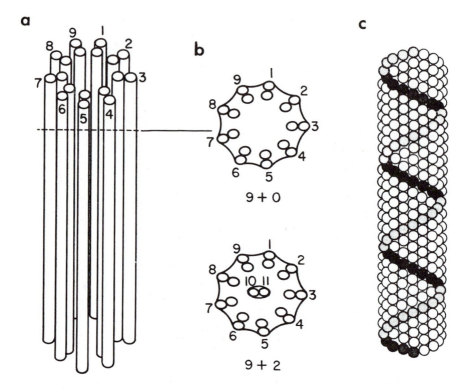

Figure 6.10 Schematized model of *(a)* flagellum (holds also for cilia and centriole); *(b)* cross section showing 9 + 0 and 9 + 2 microtubule doublets; and *(c)* the protein (flagellin) wound as an α-helix structure.

RECEPTORS

Cells possess receptors to detect changes in their environment. As cells evolved, various kinds of photoreceptors developed for phototactic movement—chloroplasts for photosynthesis and the retinal photoreceptors of the eye for visual excitation. The origin of these photoreceptors may well lie with the cell membrane itself and with the cilia and flagella. This possibility will become more apparent as we discuss the structure of these specialized photoreceptors in relation to photosynthesis (chap. 8) and to vision (chap. 10-13), as well as other photobiological processes.

REMARKS

Cells are organisms, confined within a cell membrane, possessing structural order and mobility. The cell's interior, the cytoplasm, is a very complex system of fluids and solids, but we know little of these states of matter in a living cell. In eukaryotes there are membraneous structures, *organelles,* that perform specific functions for the cell. Though considerable biochemical and structural advances have been made toward our understanding of how cells function (De Duve, 1984), we have not yet achieved a complete molecular description of even the simplest bacterial cell. As we learn more about plant and animal cells at the molecular level, we can see a basic chemical and structural pattern emerging for all living cells.

It is in search of how cells sense their environment, particularly light, that draws our interest. Let us then turn to the photoreceptor molecules and how they are chemically structured to capture the energy of solar radiation.

7

The Receptor Molecules That Do the Work

It is a common biological conception that the occurrence of pigment in animals and plants bears a clear relation to biological effects of light.
Jacques Loeb, *Dynamics of Living Matter,* 1906.

How have the various photochemical reactions involved in life processes come about? All living organisms from bacteria to humans respond in a variety of ways to solar radiation. This response is seen in behavior; a plant or animal will bend, move or swim toward or away from a light source. These phenomena are *phototropism* and *phototaxis.* Plants are able to utilize directly the visible part of the sun's radiation in the process of *photosynthesis,* converting light energy to chemical energy, which the plant then uses to manufacture sugars and organic matter. In animals, various kinds of photosensory cells and photoreceptors evolved, giving rise to the eye and to *vision.*

Many other photobiological phenomena are now recognized, such as *photoperiodism,* set by the length of day, and *photomorphogenesis,* both of which control many developmental processes such as growth, hormonal stimulation, sexual cycles of plants and animals, the timing of the flowering of plants, and the color changes in the skin of animals. Other photobiological phenomena include *photoreactivation,* the "healing" of ultraviolet damage by visible radiation, and *photodynamic action,* the photosensitization produced in the absorption of light by a molecule that becomes activated and causes destructive photo-oxidation in cells. All of these photobehavioral responses are driven by the energy of light.

What is the basis for this photosensitivity in living organisms? When the electromagnetic energies in the visible spectrum are captured by living cells, these energies are both selected and changed. The process of selection is *filtering* and the process of conversion is *transduction*. As an example, only certain wavelengths of light will be captured by specific receptor molecules.

The basis for photosensitivity in living organisms is a photoreceptor that may be as simple as a single molecule or as complex as the whole human retina. Only light absorbed by the receptor molecules in cells is effective in promoting photobiological reactions. The receptor molecules are pigments that form a complex with specific proteins within the cell membrane or the photoreceptor membrane.

These pigments, or systems of pigments, act both as electromagnetic filters by absorbing light of selected wavelengths and as transducers by converting the light energy into chemical, mechanical, or electrical energy via the cell. To better understand how such pigments function, it is necessary to identify the pigment molecules and their chemical structure and properties, before exploring the various photoprocesses in which they participate.

ACTION SPECTRA AND ABSORPTION SPECTRA

How do we identify the receptor molecules responsible for photobehavior? One means is the behavioral action spectrum. The action spectrum for a photoprocess is determined by measuring the response a microorganism, plant, or animal makes to a light stimulus of a given intensity and wavelength. The spectrum so obtained for the behavior should correspond to the absorption spectrum of the molecule or molecules responsible for this behavior. The action spectrum can then be compared to the absorption spectrum obtained by microspectrophotometry, providing that the cell and photoreceptors can be resolved microscopically.

To precisely identify a pigment with a photoprocess requires careful extraction from the cells and complete purification, the methodology of which is not always simple. The pigments are usually extracted from living organisms with various solvents and further purified by physical and chemical methods. The purified pigment extracts dissolved in pure solvents are then identified by spectroscopy. All absorption spectral data are expressed as Absorbance versus Wavelength. The Absorbance $A = \log_{10} I_0/I$, where I_0 is the intensity of the entering wavelength of light and I is the intensity of the wavelength of light transmitted through the pigment solution. The absorption spectra of the purified pigment is of great informational value in establishing its identity and molecular structure.

PIGMENTS

The pigment molecules that are vitally important for life are the chlorophylls in photosynthesis, the hemes in hemoglobin of the red blood cells, and the cytochromes, which are the respiratory enzymes that carry oxygen to the cells. These pigment molecules have a common structure, which consists of four pyrrole rings linked together forming a cyclic tetrapyrrole structure called a porphyrin. The hemes and cytochromes are iron porphyrins. Chlorophyll is a magnesium porphyrin. The metal porphyrins are catalysts that are ideal for electron transfer into photo and chemical reactions and are essential molecules for sustaining life on earth.

The porphyrins made their appearance during the evolution of organic matter and were catalysts for chemical reactions long before they became incorporated into or were synthesized by living cells. We know of this function because free oxygen is necessary for the synthesis of porphyrins. Thus they must have arisen during the transitional period when the earth's atmosphere began to be enriched with oxygen. The biosynthetic scheme (as developed by Granick, 1948, 1950, 1958, and Shemin, 1948, 1955, 1956) shows certain steps in the present-day pathways for the biosynthesis of porphyrins, chlorophyll, and heme pigments (Fig. 7.1).

Once protoporphyrin IX was formed by a slow and random series of reactions in primitive organisms, it served to increase the probability that the earlier steps would continue to occur. That this process, or something similar to it, became incorporated into living systems relatively early in evolutionary history is suggested by the universality of pyrrole in the organic world.

Pyrrole ring

The iron porphyrin, heme, is synthesized from the percursor protoporphyrin IX, which is also the precursor molecule for chlorophyll and cytochrome. The heme molecule is nearly planar; the property of planarity is due to the many double bonds in the molecule. The formula for the heme molecule is $C_{34}H_{32}O_4N_4Fe$. The iron atom forms bonds with the four nitrogen atoms of the tetrapyrrole. Heme with an attached oxygen is responsible for the red color of oxygenated blood. Four hemes that are bound to the protein globin form hemoglobin, which has a molecular weight of about 68,000. Hemoglobin is the oxygen carrier of most vertebrate red blood cells and takes up

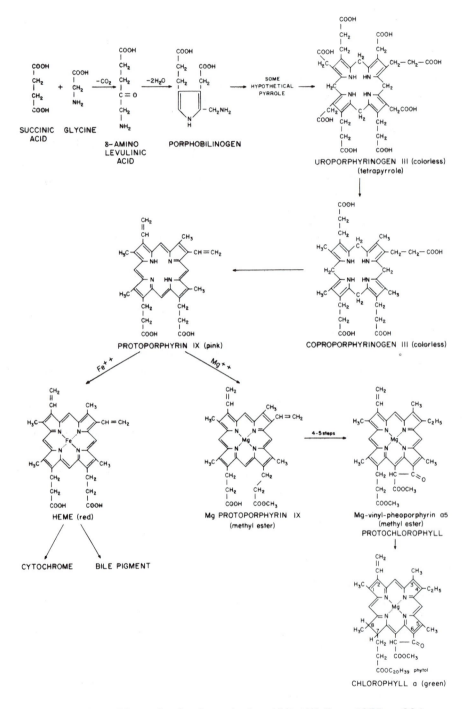

Figure 7.1 Biosynthesis of porphyrins. *(After Wolken, 1975, p. 29.)*

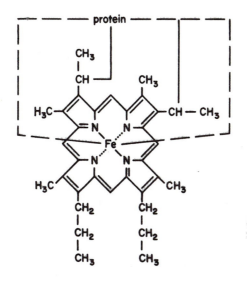

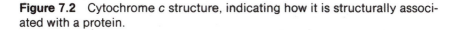

Cytochrome c

Figure 7.2 Cytochrome *c* structure, indicating how it is structurally associated with a protein.

oxygen to form oxyhemoglobin. The proportion of oxygen taken up by the hemoglobin depends on the concentration of the oxygen, or oxygen tension. Although hemoglobin is more characteristic of vertebrates, it has been found in a number of invertebrates as erythrocruorins. In other vertebrates, the blood pigment is chlorocruorin, a green pigment that contains copper and is found in many crustaceans and molluscs. All the respiratory pigments have the common property of functioning as oxygen carriers.

The cytochromes are pigmented heme proteins that carry an iron atom in an attached chemical group (Fig. 7.2). Their red color is derived from their chromophore, or prosthetic group, which is a derivative of iron protoporphyrin IX. The cytochromes function as electron carriers during the initial reactions of the photochemical processes. The cytochromes are designated by the letters *a, b, c,* and *f* and are distinguished on the basis of their spectral absorption peaks in the reduced state. For example, the spectrum of *Euglena* cytochrome-552 (a cytochrome-*c*) in the reduced and oxidized states is shown in Figure 7.3.

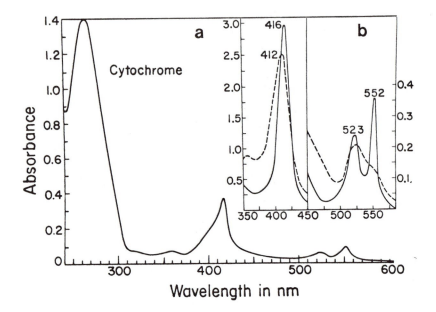

Figure 7.3 Cytochrome *c* absorption spectrum of the reduced state *(a)*. Insert *(b)* expanded to show the major reduced absorption peaks (——) and oxidized absorption peaks (- - - -). Isolated from light-grown *Euglena gracilis*. *(After Wolken, 1967.)*

Cytochrome *c* has been isolated from animals, higher plants, algae, and bacteria. The evolutionary history of cytochrome *c* is well documented and a pattern of 1.5 billion years of mutations separating yeast from man can be deduced (Fitch and Margoliash, 1967). In fact, because the rate of evolution of this molecule seems to have been relatively constant (1% change per 20 million years), cytochrome *c* can be used as a paleontological clock to time the divergence of animal species from a common evolutionary pathway (Dickerson, 1971).

Chlorophyll

More than a century ago it was recognized that chlorophyll, synthesized by all green plants, is directly associated with photosynthesis. But the chemical structure of chlorophyll was not definitely established until 1913 by Willstäter and Stoll, and then later by Fischer and Stern (1940). More recently Woodward (1961) successfully synthesized chlorophyll *a* in the laboratory.

The structure of chlorophyll can be described as tadpolelike, with a magnesium porphyrin head, in the form of a cyclic tetrapyrrole, and a phytol

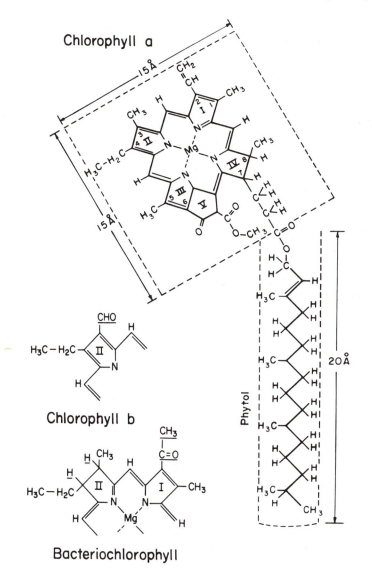

Figure 7.4 Structure of Chlorophyll *a*, Chlorophyll *b*, and Bacteriochlorophyll.

tail (Fig. 7.4). Phytol ($C_{20}H_{39}OH$) is a long chain alcohol, related to the carotenoids. The empirical formula for chlorophyll is $C_{55}H_{12}O_5N_4Mg$, and its greenness comes from the magnesium atom in the tetrapyrrole of the molecule. Chlorophyll exists in many isomeric forms; for example, chlorophyll *a* differs from chlorophyll *b* by possessing a methyl ($-CH_3$) group at the third carbon, whereas in chlorophyll *b* a formyl (=CHO) group occupies this position; chlorophyll *b* is therefore an aldehyde of chlorophyll *a*.

Chlorophyll *a* and *b* also differ in their absorption spectra (Fig. 7.5) as well as in their solubility. Chlorophyll *a* is more soluble in petroleum ether while chlorophyll *b* is more soluble in methyl alcohol. These differences in solubility make it possible to separate these two chlorophylls. Chlorophyll *a* is present in all green plants, while chlorophyll *b*, together with chlorophyll *a*, is found in all higher plants, ferns, mosses, green algae, and euglenoids. The other chlorophyll isomers, *c, d,* and *e,* are found in diatoms, brown algae, dinoflagellates, crytomonads, and crysomonads. Chlorophyll *c,* which lacks a phytol group, is soluble in aqueous alcohol. Chlorophyll *d* is believed to be an oxidation product of chlorophyll *a* in which the vinyl group at position 2 is oxidized to a formyl group, and is found together with chlorophyll *a* in most red algae. Chlorophyll *e,* together with chlorophyll *a,* is present in small amounts in yellow-green algae.

Pheophytin *a* and *b* are degradation products of chlorophyll *a* and *b* respectively, from which the magnesium has been removed. Pheophytin can be found in vivo, and is believed to have a function in photosynthetic cells.

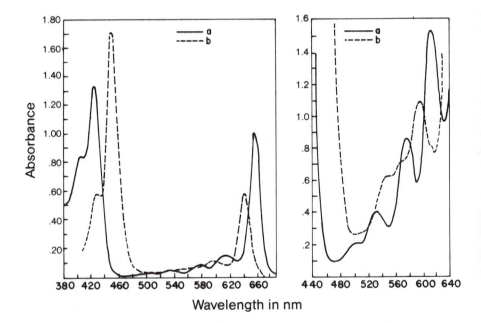

Figure 7.5 Absorption spectrum of pure Chlorophyll *a* and *b* in ethyl ether; enlargement of the spectrum between 440 nm and 640 nm.

Pheophytin is produced by acidifying chlorophyll extracts. Its formation can be observed by a color change from green to brownish yellow. Pheophytin *a* in ether has absorption peaks around 408, 505, 534, and 510 nm, accompanied by the decrease in density of chlorophyll, which has a main peak in the red near 663 nm and a small peak near 667 nm. The reverse process of inserting magnesium into pheophytin to form chlorophyll is difficult to perform in the laboratory.

Bacteriochlorophyll is found in photosynthetic purple bacteria. Bacteriochlorophyll differs from chlorophyll *a* in that the vinyl group at position 2 is replaced by an acetyl group, and in that it contains two extra hydrogen atoms at positions 3 and 4 (Fig. 7.4).

Seeds and etiolated plants (seedlings sprouted in darkness) are sometimes faintly green, although they contain no chlorophyll. Upon exposure to light, they will turn green. The substance responsible for this reaction is protochlorophyll (Fig. 7.1), the chlorophyll precursor. Protochlorophyll differs from chlorophyll in that it lacks two hydrogen atoms at positions 7 and 8 in the porphyrin part of the molecule. Thus it is an oxidation product of chlorophyll *a*.

The question of how chlorophyll, together with accessory pigment molecules within the chloroplast membranes, carries on the process of photosynthesis is discussed in chapter 8.

The Bilin Pigments and Phytochrome

The bilin pigments are so named because they were first discovered in bile. They may in fact be products of metabolic transformation of hemoglobin, other hematim compounds, and chlorophyll. The name *phycobilin* indicates that they are derivable from algae. In red and green algae there are blue pigments, phycocyanins, and red pigments, phycoerythrins. Phycocyanins and phycoerythrins consist of a chromophore and protein. Because of the similarity of the chromosphore they have been termed *phycobilins* or biloproteins. The phycobiloprotein pigments, unlike the chlorophylls and carotenoids, are water soluble and are identified by their absorption spectral peaks.

In the red and green algae, the biloproteins can utilize absorbed light energy by transferring it to chlorophyll *a* in the process of photosynthesis with an efficiency equivalent to or greater than that of chlorophyll alone. One role suggested for these pigments is that they make a larger part of the spectrum of visible energy available for photosynthesis.

It is interesting to note that the phycobiloprotein chemical structure, like that of chlorophyll, is a tetrapyrrole, but with an open ring porphyrin arranged linearly (Fig. 7.6). For example, the structure of phycocyanin closely resembles that of phytochrome (Fig. 7.7), the pigment responsible for

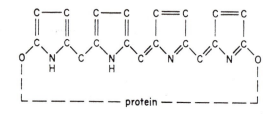

Phycobiliprotein

Figure 7.6   Phycobiloprotein basic structure. Note similarity to phytochrome, Figure 7.7.

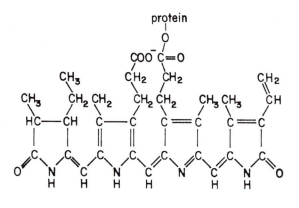

Phytochrome

Figure 7.7   Phytochrome structure. (See Figs. 13.6 and 13.7.)

plant photoperiodism. The phycocyanins and phytochrome structures with their system of conjugated double bonds also resemble the carotenoid structure (Fig. 7.8).

The Carotenoids

Of the major pigment groups found in nature, carotenoids are perhaps the most abundant of all pigments. They are easily synthesized by bacteria, algae, fungi, and plants. The carotenoids are all shades of yellow, orange, and red and are widely distributed throughout all plants. In green plants carotenoids are usually found together with the chlorophylls. The carotenoids possess 40 carbon atoms in each molecule (Fig. 7.8). They are named for

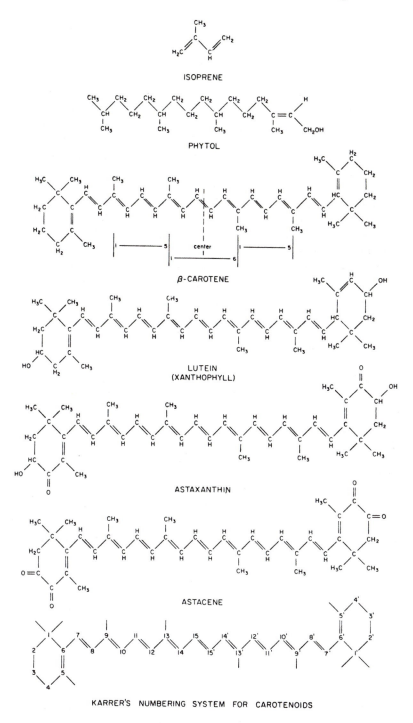

Figure 7.8 Structures of various carotenoids.

their most familiar substance, carotene, and are divided into two main groups: the carotenes (pure hydrocarbons), the most abundant of which is all-*trans*-β-carotene, $C_{40}H_{56}$ and the xanthophylls, $C_{40}H_{56}O_2$ (oxygen-containing derivatives). The oxygen atoms can be in hydroxyl, carboxyl, or methoxyl groups. From their structure, carotenoids can be considered to be built from isoprene units. In their linear arrangement a carotenoid molecule consists of four radicals of isoprene (2-methyl-1, 3-butadiene) residues. The isoprene units are linked so that the two methyl groups nearest the center of the molecule are in position 1 and 6, while all other lateral methyl groups are in positions 1 and 5.

The carotenoid molecule is made up of a chromophoric system of alternating single and double interatomic linkages between the carbon atoms, called a polyene chain of conjugated double bonds. The large number of these conjugated double bonds offers the possibility of either *cis*- or *trans*-geometric configurations. It is estimated that there are about 20 possible geometric isomers of β-carotene, of which six *cis*- isomers have been discovered in nature. The spectral characteristics, and therefore the color of the carotenoid, are largely determined by the number of conjugated double bonds in the molecule.

The biosynthesis of carotenoids (Fig. 7.9) is generally associated with the 20-carbon, aliphatic alcohol phytol, which is the colorless part of the ester-comprising chlorophyll. The striking resemblance between the carotenoid skeleton and phytol holds also for the details or spatial configuration.

Though carotenoids are easily and abundantly synthesized by plants, multicellular animals cannot synthesize carotenoids but must obtain them by ingesting plants. Animals can modify and even degrade the carotenoid molecules to serve their special needs. For example, animals convert β-carotene ($C_{40}H_{56}$) to vitamin A ($C_{19}H_{27}CH_2OH$), which is necessary for life, and whose aldehyde, retinal, is necessary for vision. The difference in their absorption spectra are shown in Figure 7.10.

An important change occurred during the evolutionary development of animals; that is, they became dependent on the ingestion of plants as their source of carotenoids. It is not the ingested plant carotenoids themselves but their degraded derivative, vitamin A, that is necessary for all animal life (Fig. 7.11). Thus the carotenoids play a central role in the biochemical evolution from the plant C_{40} (β-carotene) to animal C_{20} (vitamin A). The importance of this change to the evolution of the eye and to an understanding of the photochemistry of vision will be elaborated in chapter 12.

Flavins

Riboflavin (vitamin B_2) is taken here as an example of the flavins and flavoproteins. Riboflavin is synthesized by a number of microorganisms and by most higher plants, particularly in growing leaves. It occurs in small

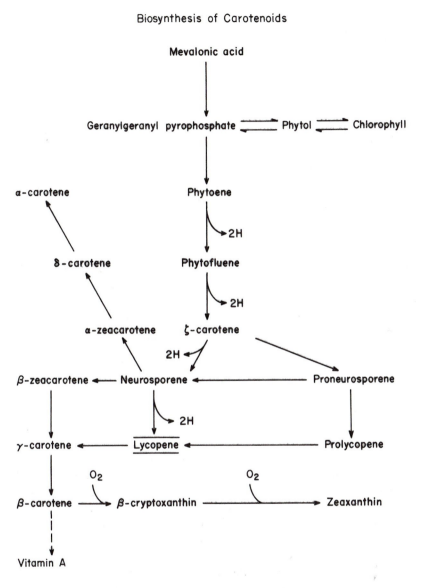

Biosynthesis of Carotenoids

Figure 7.9 The biosynthesis of carotenoids.

concentrations in nearly all animal tissues. As one of the B vitamins, it is a factor in cell respiration.

The structure of the riboflavin molecule consists of a tricyclic carbon-nitrogen ring system, isoalloazine, with methyls at the 7 and 8 position and a ribose sugar attached to the 10-nitrogen (Fig. 7.12). Riboflavin in solution is yellow and has spectral absorption peaks in the oxidized state around

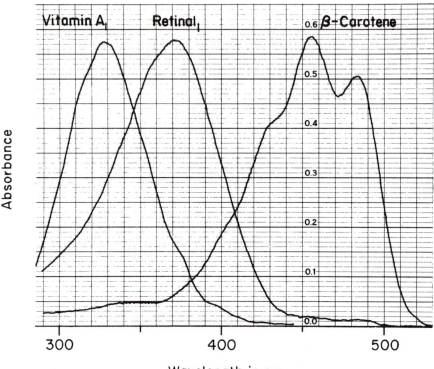

Figure 7.10   Absorption spectra of β-carotene, retinal, and Vitamin A$_1$.

221-227, 265-270, 365-370, and 445-460 nm (Fig. 7.13). One of the more important chemical properties of riboflavin is its ability to change reversibly from the yellow-colored, oxidized form to the colorless, reduced form. Upon ultraviolet excitation it fluoresces green about 520-560 nm. Riboflavin is light-sensitive in neutral or acid solution and forms lumichrome with absorption peaks near 223, 260, and 360 nm.

Riboflavin mononucleotide is riboflavin-5'-phosphate, and is usually referred to as FMN. Warburg and Christian (1938a, 1938b, 1938c) discovered that an enzyme catalyzing the oxidation of D-amino acids, D-amino oxidase, was a flavoprotein containing a prosthetic group distinct from FMN. The structure of this coenzyme, flavin adenine dinucleotide, is called FAD.

Riboflavin and the flavoproteins may be significant molecules in photoreceptor processes associated with phototactic behavior in algae, protozoa, fungi, plants, and animals. Flavoproteins may even be involved in the biochemistry of the visual process (Fox, 1960). Riboflavin is found in the eye, particularly in the retina of certain mammals, in the pigment epithelium

Vitamin A₁

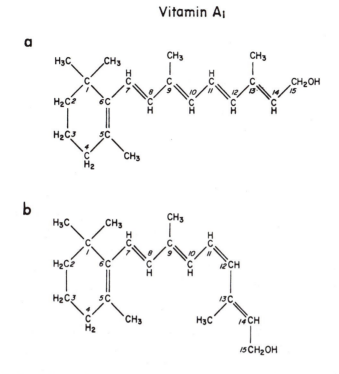

Figure 7.11 Vitamin A₁ geometric isomers, structures of *(a)* all-*trans* and *(b)* 11-*cis*.

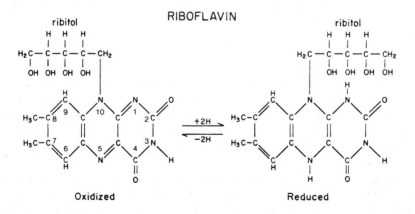

Figure 7.12 Riboflavin (Vitamin B₂) structure in oxidized and reduced states.

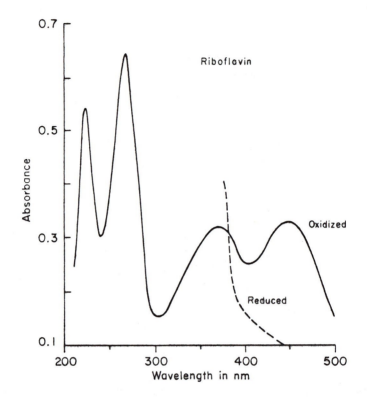

Figure 7.13.   Riboflavin absorption spectrum in oxidized and reduced states. Reduced by dithionite.

of fish, and between the pigment epithelium and the choroid of frogs, rabbits, rats, and cattle.

Melanins

The most abundant pigments found in vertebrates are the melanins. The nature of the pigment is determined by the genetic makeup of the organisms and the biological function. These functions include camouflage from predators, sexual recognition within a species, and sexual attraction. Difference in pigmentation varies with exposure to solar radiation and the aging process of the organism.

Melanin pigments are black, brown, yellow, or red and occur in all animal phyla. They are found in hair, skin, and eyes. The main function of melanin in the skin is protection against solar ultraviolet radiation. The biosynthesis of melanin can be observed simply by watching how a cut apple, potato, or banana turns brown-black upon exposure to air.

Melanin is a biochrome of high molecular weight and can be produced in vitro by oxidation of the amino acid tyrosine with the enzyme tyrosinase, a copper-containing protein. The first step in this reaction is the production of dopa (2,3-dihydroxyphenylalanine), which is then oxidized enzymatically to dopa-quinone. After this stage there occurs a complicated series of further oxidations and polymerizations leading to the formation of tyrosine-melanin.

Very little is known of the structure of natural melanin produced by the chromatophores, the melanophores, and the melanocytes of cells. The problem is further complicated by the fact that melanin is always bound to a protein. However, all available evidence indicates that tyrosine is the precursor of melanin and that the biosynthesis follows the pathway of tyrosine→dopa→melanin.

REMARKS

In our discussion of the molecules for life we mentioned that only a few atoms make up the protein, lipid, and carbohydrate molecules. Similarly in the pigment molecules just described only a few atoms make up these molecular structures, and a surprisingly small number of such molecules actually work in the photoprocesses of living organisms.

One asks why this list of specialized molecules, for photochemistry, is so relevant for biology when there is an unlimited variety of molecules that function without light for biochemistry. The answer lies in their molecular structure of long-chain, conjugated carbon-to-carbon bonds ($-C=C-C=C-$). We find them in carotenes, in linear pyrolles and ring structures, and in the porphyrins and flavins, which are so well designed to capture the visible bands of energy. It is more than interesting to note that the absorption spectra of the pigments just described cover the sun's spectrum of radiation reaching the surface of the earth and that their absorption spectra all fall in the visible range cluster about the solar energy peak that strikes the earth around 500 nm.

With this information on pigments, their structure and absorption characteristics, we can begin to see how these pigment molecules are the photoreceptors for photosynthesis, vision, and other photoprocesses of living cells that are driven by the energies of light.

8 *Photosynthesis*

From the simplest substances, carbon dioxide, water,
and sunlight, autotrophic plants produce enormous
quantities of organic matter. . . . Synthesis of all this
diverse vegetable material hinges upon photochemical
reactions that take place within the green parts
of plants.
H. H. Strain, *Ann. Rev. Biochem. 13:* 591, 1944.

Of major importance to life is photosynthesis, a mechanism that evolved enabling organisms to utilize solar energy directly. In the process of photosynthesis the energy of light is converted to chemical energy via chlorophyll (Fig. 8.1) in the synthesis of organic compounds, such as sugars and starches, according to the fundamental equation:

$$CO_2 + H_2O \xrightarrow[\text{chlorophyll}]{\text{light}} \underset{\substack{\text{(Organic} \\ \text{compounds)}}}{(CH_2O)} + O_2.$$

Photosynthesis is ultimately the source of food for all animal life and a vital process for sustaining life on Earth.

Photosynthetic organisms have existed on Earth for at least 2×10^9 years. The total amount of organic compounds formed each year by photosynthesis is about 1.6×10^{17} grams or 100 billion tons. The total mass of organic material produced by green plants during the biological history of the earth has been estimated to be about 6×10^{25} grams. This total is an enormous weight when compared to the mass of the earth, which is 6×10^{27} grams, as it represents 1% of Earth's mass.

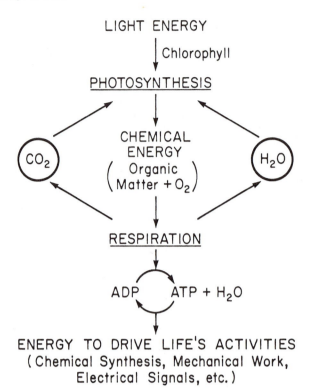

LIGHT ENERGY

Chlorophyll

PHOTOSYNTHESIS

CHEMICAL
ENERGY
$\left(\begin{array}{c}\text{Organic}\\\text{Matter} + O_2\end{array}\right)$

CO_2 H_2O

RESPIRATION

ADP   ATP + H_2O

ENERGY TO DRIVE LIFE'S ACTIVITIES
(Chemical Synthesis, Mechanical Work,
Electrical Signals, etc.)

Figure 8.1 Schematic energy relationships between photosynthesis and respiration.

A biochemical operation of this magnitude would quickly deplete the earth's atmosphere of its carbon dioxide. The fact that our atmosphere still contains CO_2 means that it is being returned to the atmosphere by equally large-scale processes. That is, the rate of CO_2 consumption in photosynthesis is just about balanced by the rate of restoration. This balance is observed in Figure 8.2, in which the spectrum for the uptake of CO_2 in photosynthesis corresponds to the absorption spectral peaks for chlorophyll. Respiration by plants and animals is responsible for this restoration of atmospheric CO_2 in accordance with the equation:

$$(CH_2O) + O_2 \longrightarrow CO_2 + H_2O.$$
(Organic
compounds)

This restoration is supplemented by the decay of organic matter and the burning of fuels, both of which yield CO_2. As a result, the average surface concentration of CO_2 in the air, about 0.03%, has remained practically con-

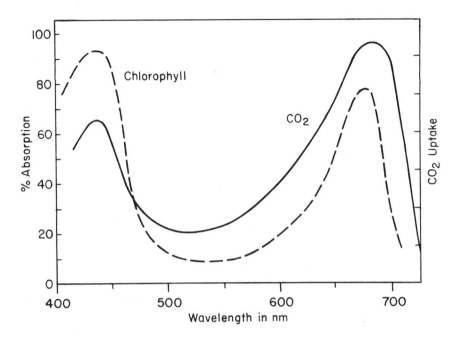

Figure 8.2 Action spectrum for relative effectiveness for CO_2 uptake in photosynthesis compared to absorption spectral peaks for chlorophyll.

stant for thousands of years. The set of processes maintaining the CO_2 balance (the carbon cycle) is important not only in maintaining a constant chemical atmospheric environment but also in regulating the surface temperature of the earth.

Photosynthesis is equally important in that it regulates the oxygen content of the atmosphere. If too much oxygen had been created too early, the greenhouse effect would have been lost and the global temperature would have plunged. If oxygen production proceeded too slowly, the greenhouse effect would have trapped too much heat. Therefore, biological and other mechanisms were at work to modulate the atmospheric oxygen, because there was very little, if any, oxygen in the atmosphere until the first photosynthesizing organisms evolved. It has been estimated that the current rate of photosynthesis produces almost the entire oxygen content of the atmosphere necessary for the respiration of all animals, near 20%. The oxygen concentration has been maintained at its present level for a long period of time as a result of a state of balance in the oxygen cycle. The emergence of photosynthesis then has played a crucial role in the evolution of other forms of life on Earth.

The beginning of our understanding of photosynthesis goes back to the

1770 s, when Lavoisier (1774) in France determined the composition of air and Joseph Priestly (1772) in England discovered oxygen. Priestly then showed that oxygen was produced by algae and green plants. The Dutch physician Jan Ingen-Housz observed in 1779 that in the process of plant respiration green plants at night or in the dark gave of a "dangerous air," carbon dioxide, which was purified by sunlight. The Swiss clergyman Jean Senebier in 1782 was able to demonstrate that the carbon dioxide was absorbed by the leaves of green plants and when exposed to sunlight, oxygen evolved. In 1798 Ingen-Housz published his theory that carbon, already recognized at the time as an important element in the composition of organic molecules, was derived from carbon dioxide during photosynthesis. It was another Swiss scientist, Nicholas de Saussure (1804), who surmised that water played an essential role, and the picture changed to one in which light acted on both carbon dioxide and water. Pelletier and Caventou in 1818 identified chlorophyll as the green pigment of plant tissue, and by 1837 chlorophyll was identified with the chloroplasts of plant cells. However, the crux of the phenomenon of photosynthesis was visualized in 1845 by Robert Mayer, a German physicist and physician, who pointed out that the photosynthetic process was the conversion of light energy to chemical energy. By 1882, Englemann demonstrated that the site of photosynthesis resided within the chloroplast of plant cells and that upon light absorption oxygen was liberated. From then on, meaningful research to understand the mechanism of photosynthesis had begun. An account of these early investigations into photosynthesis as it developed is summarized by Rabinowitch in the three volumes of *Photosynthesis and Related Processes* (1945, 1951, 1956).

COMPARATIVE ASPECTS OF PHOTOSYNTHESIS

Photosynthesis evolved in early anaerobic bacteria. These bacteria were able to utilize inorganic compounds such as hydrogen sulfide and hydrogen gas, as well as organic compounds in their environment. In time, some of them developed metabolic pathways that led to the synthesis of porphyrins, bacterial chlorophyll, and hence to the evolution of bacterial photosynthesis. When photosynthetic bacteria became well established, a second kind of photosynthesis became possible, one in which the uptake of hydrogen was accomplished by the photolysis of water molecules. As a result, free oxygen entered the atmosphere and became available for further chemical synthesis.

For anaerobic bacteria oxygen was a deadly poison, but as the population capable of photosynthesis increased so did the oxygen content of the oceans and the atmosphere. It has been hypothesized that 3 billion years ago the level of atmospheric oxygen was less than 0.001 of the present and by 1 billion years ago had probably increased to about 0.1 of the present.

The first organisms on earth that adapted to the presence of free oxygen led to the appearance of blue-green algae now known as cyanobacteria. Compared with algae and green plants, these bacteria are primitive. The success of the cyanobacteria was manifest in the evolution of new kinds of bacteria that utilized free oxygen in their metabolic processes. Van Niel (1941, 1943) pointed out that the photosynthetic cyanobacteria and the purple bacteria represent remnants of an originally much wider class of organisms having a photosynthetic system simpler than that of green plants. The metabolism of purple bacteria could serve as an example for the kind of photochemistry that may have preceded that of the green plants on the evolutionary time scale. These organisms cannot evolve oxygen, though many tolerate oxygen. Because they require energy-rich hydrogen donors to reduce CO_2, they do not contribute much to the store of free energy in the living world. In photosynthetic bacteria, bacteriochlorophyll is responsible for the utilization of light energy and is similar in structure to the chlorophyll of green plants (Fig. 7.4).

Let us briefly examine some aspects of the comparative biochemistry of photosynthesis. Organisms are classified as: *autotrophic* (those which obtain their energy for growth from sources other than organic molecules), *chemoautotrophic* (those which obtain their energy from oxidizable inorganic chemicals), and *photoautotrophic* (those which obtain their energy directly from light). The photoautotrophs fall into three separate groups: green plants, pigmented sulfur bacteria, and pigmented nonsulfur bacteria. Their light-driven reactions can be expressed by the following equations:

$$\text{Green plants: } CO_2 + H_2O \xrightarrow{\text{light}} \underset{\text{Organic matter}}{(CH_2O)} + O_2$$

$$\text{Sulfur bacteria: } CO_2 + H_2S \xrightarrow{\text{light}} (CH_2O) + S$$

$$\text{Nonsulfur bacteria: } CO_2 + \text{succinate} \xrightarrow{\text{light}} (CH_2O) + \text{fumarate.}$$

For the sulfur bacteria, H_2S can be replaced by $Na_2S_2O_3$, $Na_2S_4O_7 \rightarrow H_2SO_4$, or H_2Se. For nonsulfur bacteria, the organic donor, succinate, can be replaced by many different organic acids that have two electrons to spare. This replacement is schematically generalized by

$$CO_2 + H_2A \xrightarrow{\text{light}} (CH_2O) + A.$$

The characterization of the various types of photosynthesis led to a generalization by Van Niel (1941, 1949) of the "comparative biochemistry of photosynthesis." Actually, the photosynthetic reaction can be schematized even further without involving carbon dioxide, for in the Hill reaction (production of O_2 by isolated chloroplasts in light), a quinone or ferric ion can accept the hydrogens that are activated by the light reaction, so the general formula can be written either:

$$B + H_2A \xrightarrow{\text{light}} BH_2 + A$$

or simply as a light-induced oxidation-reduction reaction.

Many researchers attacking the mechanisms of photosynthesis hypothesized that the photochemical reaction in green plant photosynthesis is a photolysis (light-induced decomposition) of water. Indeed, experimental evidence has shown that water is photolized in both plant and bacterial photosynthesis. Therefore, a more acceptable equation for the bacterial systems would include both water and a general hydrogen donor and would also exclude oxygen evolution, for example:

$$CO_2 + 2H_2A + H_2O \xrightarrow{\text{light}} (CH_2O) + 2H_2O + 2A$$

A simple formulation for the mechanism of the overall process has yet to be conceived. Some aspects, however, are reasonably well understood. The general photosynthetic light reaction can be represented schematically as:

$$HOH \xrightarrow[\text{organism}]{\text{light}} \begin{cases} [H] \xrightarrow{\text{electron transport carriers}} (CH_2O) \\ [OH] \xrightarrow[\text{organic peroxides}]{\text{via } O_2 \text{ carriers}} H_2O + O_2 \text{ or oxidized material} \end{cases}$$

The photolysis of water can be regarded then as the major achievement of the chlorophyll system. Everything else follows via reaction patterns that have become well known in general biochemistry.

THE MECHANISM

In examining the photosynthetic mechanism there are two distinct and related processes that occur. One process is the conversion of carbon dioxide to carbohydrates indicated by the equation for green plants. The other process is the conversion of light energy to chemical energy, which is the more difficult to understand. Thus, photosynthesis is a series of light and dark reactions. The reduction of carbon dioxide to carbohydrates is a dark reaction and is separate from the primary light quantum conversion.

We can say, then, that light energy is converted into chemical energy to form carbohydrates and oxygen. In this process the light energy is absorbed by chlorophyll and related pigments and is converted into chemical potential energy in the form of certain compounds. These compounds then react with water, liberating oxygen, reducing agents, and other cofactors that also contain high chemical potential energy. Finally, these reducing and energetic cofactors react with carbon dioxide and other inorganic compounds to produce organic compounds.

Melvin Calvin and his associates, in the 1940s at the University of California at Berkeley, began to study the pathway of carbon reduction during photosynthesis using ^{14}C. They identified phosphoglyceric acid as the first stable product of carbon reduction during photosynthesis. With the development of two-dimensional paper chromatography combined with radioautography, analytical tools were available for separating minute amounts of radioactive compounds formed in the plant during photosynthesis. By these methods the intermediates in the carbon reduction cycle were found to be sugar phosphates. Calvin (1962) confirmed Blackman's hypothesis that light was necessary for only two processes: (1) to produce ribulose diphosphate, the acceptor of CO_2, by phosphorylation of ribose monophosphate, and (2) to permit the reduction of carboxyl groups of phosphoglyceric acid (PGA) in the aldehyde group by the intermediate 1,3-diphosphoglyceric acid. In these reactions the donor of the phosphoryl group is ATP (adenosine triphosphate).

ENZYMES

Researchers have looked for the participation of a specific enzyme system in photosynthesis for capturing electrons. Hill and Bendall (1960) experimentally demonstrated that a cytochrome, or a cytochrome system, is coupled to the chlorophyll-protein complex in the chloroplast, and that it functions as an electron carrier during the initial reactions of the photochemical process. As Hill and Bendall (1960) pointed out, mitochondria and chloroplasts show close resemblance with respect to the structure-bound cytochromes (Table 8.1), indicating that the chloroplasts belong to the same category as the mitochondria, with cytochrome *a* being replaced by the chlorophylls. There is also the possibility that the photochemistry initiated by light absorption in

Table 8.1 Cytochromes in Mitochondria and Chloroplasts

Mitochondria and Chloroplasts	*Cytochromes*		
Yeast mitochondria	a	b	c
Etiolated barley yellow plastids	a	b_6	f (modified c)
Green plant chloroplasts	—	b_6	f (modified c)

Source: After Hill and Bendall, 1960, and Wolken, 1975, p. 72.

photosynthesis involves the cytochrome directly. For example, in *Euglena*, two spectrally different cytochromes have been isolated. One from the light-grown photosynthetic cells is designated as cytochrome-552 (a *c*-type cytochrome). In its reduced state its absorption peaks are at 552, 523, and 416 nm (Fig. 7.3). The other cytochrome isolated from dark brown *Euglena* is referred to as cytochrome-556, because in the reduced state it has absorption peaks in the visible at 556, 525 and 412 nm. Its spectrum is close to that of cytochrome *f*, which is associated with green plants. The ratio of chlorophyll *a* to cytochrome-552 is approximately 300:1.

Similarities in the photosynthetic systems between cyanobacteria, algae, and green plants have led to the suggestion that the cyanobacteria became a symbiot of an early cell and that the chloroplast ferredoxins are therefore derived from a common ancestor. Arnon (1965) has shown that ferredoxin is also a key photochemical component of the process of photosynthesis by chloroplasts. Unlike the cytochromes that exhibit well-defined absorption peaks in the reduced state, ferredoxins have distinct absorption peaks in the oxidized state. According to Arnon, the photoreduction of ferredoxin is coupled with oxygen evolution and with photosynthetic phosphorylation. It is interesting to note that the ratio of chlorophyll to ferredoxin molecules was also found to be about 300:1.

Plastoquinone is another important intermediate in photosynthesis and is found in cyanobacteria, in green, red, and brown algae and in green plant chloroplasts. The ratio of the total various quinones to chlorophyll was found to be about 150:1 (Amesz, 1973).

TWO PHOTOSYSTEMS IN PHOTOSYNTHESIS

The first experimental observations indicating that the photosynthetic process was not a simple photoreception sensitized by chlorophyll were made by Emerson and Lewis (1943). They observed that the quantum yield in photosynthesis was reasonably constant between 500 and 680 nm, but dropped dramatically beyond 680 nm. Because chlorophyll *a* is the major light absorber in this region of the spectrum (Figs. 7.5 and 8.3), it seemed that light absorption by chlorophyll *a* alone was not sufficient for photosynthesis to proceed. Therefore, it was assumed that in the living cell there are forms of chlorophyll *a* that differ in the way they are complexed with their proteins, or perhaps are associated with accessory pigments. To understand these experimental results, two pigment photosystems for photosynthesis were postulated (Duysens, 1964).

Emerson (1956), using *Chlorella* and monochromatic light, observed that the low efficiency of photosynthesis beyond 680 nm could be considerably

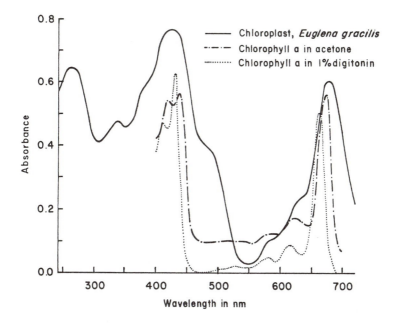

Figure 8.3 Absorption spectrum of *Euglena gracilis* chloroplast *in situ* (obtained by microspectrophotometry) (————); absorption spectra of purified chlorophyll *a* in acetone (– · – · – · –); chlorophyll *a* in 1% digitonin (· · · · · ·).

improved by simultaneous illumination with shorter wavelength blue light at 480 nm. Thus, the low efficiency of absorption in the far red beyond 680 nm would require another pigment complex absorbing below 680 nm. To account for these experimental results, two pigment photosystems for photosynthesis are required, known as Photosystem I and Photosystem II.

A scheme for this complex process is illustrated in Figure 8.4. In examining this scheme, water serves as the electron donor in a photoreaction promoted by a chlorophyll complex, P_{680}, referred to as Photosystem II. The electron acceptor is an unknown compound Q. Its redox potential is around 0.0 to +0.18 V. The reducer Q transfers its electron through a series of compounds, including the enzymes plastoquinone and several cytochromes, and finally to Photosystem I. This chlorophyll complex, containing mainly chlorophyll a absorbing at longer wavelengths, is called P_{700}, so named because it behaves differently from a typical chlorophyll, and it has a redox potential of +0.4 V. Thus, an electron moving between the two photosystems loses the equivalent to about 0.2 to 0.4 V. This amount of energy is enough to promote the formation of one or two ATP molecules from ADP and inorganic phosphate. In Photosystem I the electron acceptor is an unknown compound *(X)*. The light absorbed by Photosystem I is then used to reduce

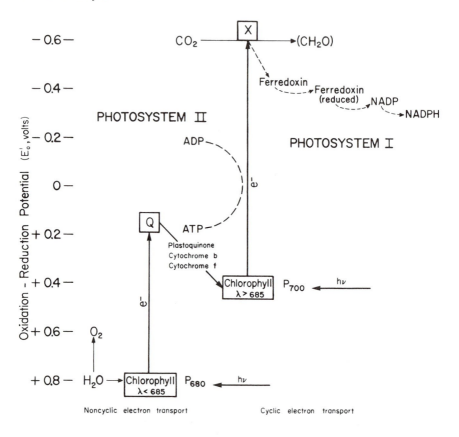

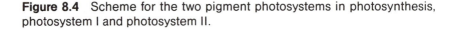

Two Photosystems in Photosynthesis

Figure 8.4 Scheme for the two pigment photosystems in photosynthesis, photosystem I and photosystem II.

ferredoxin. The final product of this electron transport scheme is NADP (nicotinamide adenine dinucleotide phosphate).

In summary, the two photochemical oxidation-reduction reactions are driven by two pigment systems, Photosystem I and Photosystem II. The reduced electron acceptor of Photosystem I reduces $NADP^+$ via ferredoxin; the oxidized electron donors of Photosystem II produce oxygen by oxidation of water. Photosystem II contains chlorophyll *a* absorbing at relatively shorter wavelengths, and a relatively large part of the so-called accessory pigments, chlorophyll *b* (in green algae and higher plants), and phycobilins (in red algae and cyanobacteria). Therefore, Photosystem I and Photosystem

II provide the high energy phosphates (reduced NADP and ATP) needed for the synthesis of carbohydrates and proteins from CO_2 and water.

THE CHLOROPLAST IN PHOTOSYNTHESIS

The question of how the chloroplast functions as a device for energy capture and energy conversion, from light to chemical energy, has not been fully answered.

Our understanding of photosynthesis lies within the chloroplast photochemistry and molecular structure. In unraveling this chemistry and molecular structure, we gain insight into how the chloroplast functions in photosynthesis. Chemical analysis of chloroplasts isolated from a variety of plant species shows that the major constituents are protein (35-55%), lipids (18-37%), chlorophylls (5-10%), carotenoids (2%), inorganic matter (5-8% on a dry weight basis), and the nucleic acids RNA and DNA (1-3%). All chloroplasts, except for bacterial chromatophores, contain chlorophyll a, and all higher plants and green algae contain, in addition, chlorophyll b. The total number of chlorophyll molecules per chloroplast is around 1.0×10^9 (Table 8.2).

Table 8.2 Chlorophyll Concentration in the Chloroplast

Organism	Volume of Chloroplast in ml	Chlorophyll Molecules per Chloroplast	Concentration of Chlorophyll (moles/liter)
Elodea densa (green plant)	2.8×10^{-11}	1.7×10^9	0.100
Mnium (moss)	4.1×10^{-11}	1.6×10^9	0.065
Euglena gracilis (algae flagellate)	6.6×10^{-11}	1.02×10^{9a}	0.025
Poteriochromonas stipitate (crysomonad)	1.1×10^{-11}	0.11×10^9	0.016

Source: After Wolken, 1967, p. 67 and Wolken, 1975, pp. 73-75.

[a]Number of chlorophyll molecules $1.02 \times 10^9 (0.88 - 1.36 \times 10^9)$
(calculated from chloroplast extract in solution)

1.34×10^9 (calculated from a single chloroplast using microspectrophotometry)

We can now ask what function the carotenoids play in the primary photosynthetic reaction. The bacterial photosynthetic systems are unique in being anaerobic and not evolving oxygen. The bacterium *Rhodopseudomonas spheroides* pigment from the blue-green mutant lacks the carotenoids present in the wild type. This mutant has been extensively studied and found to grow well photosynthetically without these carotenoids that were considered necessary (Stanier, 1959). However, when the mutant cultures were exposed to light and oxygen, rapid death and bacteriochlorophyll decomposition occurred. This drastic change demonstrated that the light-trapping process was not dependent on the carotenoids if oxygen was absent, and indicated that bacteriochlorophyll, and hence chlorophyll, was the primary pigment necessary for the light reaction. These experiments also indicated that the carotenoids were not essential, except in catalytic amounts, for green plant photosynthesis, but were necessary for protection against photodynamic destruction. Other investigations give support to the view that the carotenoids participate directly in the primary act of photosynthesis.

Action spectra have shown a direct relationship between the synthetic pathways of chlorophyll and carotenoid synthesis in algae and plants (Wolken, 1967; Ogawa et al., 1973). The relationship between chlorophyll and carotenoid pigments is probably that the C_{20} phytol chain of chlorophyll is derived from precursors of C_{40} carotenoids. The blue-green bacteria mutant does have the C_{40} carotenoid precursor, phytoene, which can give rise to the phytol chain via a divergent pathway (Fig. 7.9). The carotenoid synthetic pathway beyond phytoene is genetically blocked in the mutant. One hypothesis is that the carotenoids combine with the oxidized portion of the photosynthetically split water molecule by forming epoxides across the numerous double bonds, with one or more epoxide groups resulting:

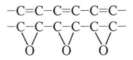

Such epoxide formation has been demonstrated in leaves and in vitro. Therefore the carotenoids can act as a filter to screen the light and prevent photo-oxidation (photodestruction) of chlorophyll at high light intensities. On the other hand they could function as an accessory pigment molecule in the energy transfer process.

The fact that DNA is present in chloroplasts indicates that they possess an autonomous genetic system different from the cell itself. The amount of DNA in chloroplasts is about the same as in *Escherichia coli,* and possesses sufficient genetic information for a large number of physiological functions. Chloroplast DNA has a nucleotide composition sharply different from that of the nuclear DNA (Brawerman and Eisenstadt, 1964).

Chloroplasts also contain messenger RNA in sufficient quantity for maximum activity of their protein-synthesizing system. A mechanism could be postulated by assuming that the messenger RNA molecules for the structural proteins of the chloroplast are generated in situ by the chloroplast DNA. The replication and turnover of chloroplast DNA of *Euglena* have been shown to be more rapid than those of nuclear DNA (Manning and Richards, 1972). This information raises many interesting questions regarding chloroplast origins and evolution (Cohen, 1970, 1973; Sager, 1972; and Margulis, 1970, 1982).

THE CHLOROPLAST STRUCTURE

How is the chloroplast structured for function in photosynthesis? The chloroplasts in photosynthetic bacteria are described as *chromatophores,* in algae as *plastids* and in all green plants as chloroplasts. The chloroplasts of green plants contain *grana* that form closed flattened vesicles or discs; these membrane structures are referred to as *thylakoids*. The chloroplasts of algae and green plants are of various shapes, but generally they are ellipsoid bodies from 1 μm to 5μm in diameter and from 1 μm to 10 μm in length. Chloroplasts observed with the polarizing microscope show both form and intrinsic birefringence. With the fluorescence microscope they show measurable fluorescence. These observations indicate that chloroplasts possess a highly ordered molecular structure.

Electron microscopy of chloroplasts in a variety of plants reveals that they consist of membranes as seen in a section through the *Euglena* chloroplast and the green plant *Elodea* chloroplast (Figs. 8.5 and 8.6). The elongated chloroplast of *Euglena* is seen to be made of regularly spaced membranes. These membranes consist of lipids and proteins, and the intermembrane spaces contain water, enzymes, and dissolved salts.

The number of chlorophyll molecules per chloroplast, from photosynthesis bacteria to higher plants, is of the order of 10^9 molecules (Table 8.2). The number of chlorophyll molecules is directly related to the number of membrane surfaces, suggesting a mode of growth regulation on the molecular level for chloroplast development. That is, chlorophyll molecules would be spread as monolayers on the surfaces of the membranes as depicted in the molecular model (Fig. 8.7), maximizing the surface area of the chlorophyll molecule for light absorption and for energy transfer at specific sites on the membrane. Such a highly ordered membrane structure not only provides for the energetic interaction of the chlorophyll and carotenoid molecules but also provides reactive sites for the necessary enzymatic reactions.

To establish that the chlorophyll molecules are spread as a monomolecular layer on the membrane surfaces, the area available for the porphyrin part of the chlorophyll molecule was calculated (Wolken, 1975). To do this, the

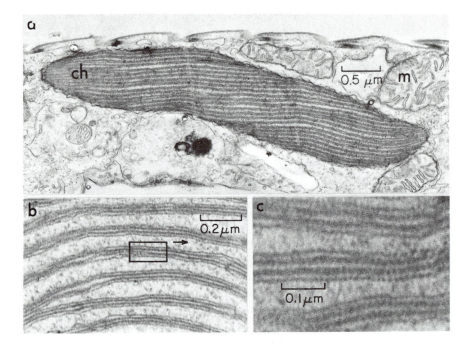

Figure 8.5 The chloroplast (ch) structure of *Euglena gracilis (a)* m, mitochondria. Note lamellae at higher magnification *(b)* and at greater resolution *(c)*. Electron micrographs.

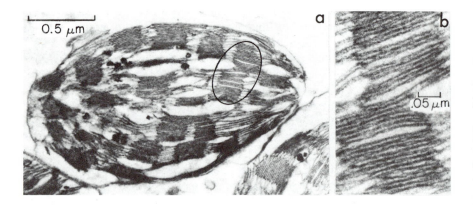

Figure 8.6 The chloroplast structure of a higher plant, *Elodea densa (a)*. Note grana enlargement in *(b)*. Electron micrographs. (Courtesy of Dr. K. Mühlethaler, Technische Hochschule, Zurich.)

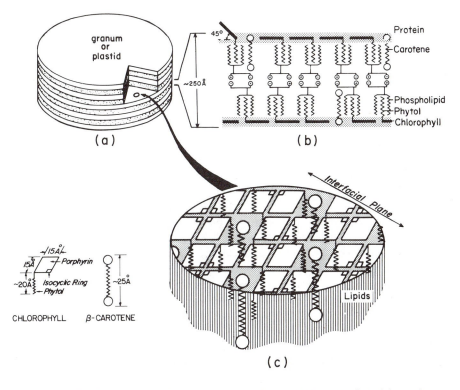

Figure 8.7 Chloroplast structural molecular model as depicted in *a, b, c. (According to Wolken, 1975, p. 83.)*

geometry of individual chloroplasts (their length, diameter, and number of membrane surfaces) was measured from numerous electron micrographs. The calculated cross-sectional area of the chlorophyll molecule was found to be 222 Å^2 for the *Euglena* chloroplasts; the cross-sectional area in a variety of plant chloroplasts was found to be around 200 Å^2. These calculations correspond well with the cross-sectional area of a porphyrin molecule when spread on a water-air interface.

So, in the model, we assume the chlorophyll molecules are oriented as a monolayer on the protein surfaces of the protein-lipid membranes. The chloroplast lamellar network shows that four chlorophyll molecules are united to form tetrads and are oriented so that only one of the phytol tails of chlorophyll is located at each intersection of the rectangular network (Fig. 8.7c). This arrangement has the advantage of leaving adequate space for at least one carotenoid molecule for every three chlorophyll molecules. Because the molecular weight of the carotenoid molecules is one-half to two-thirds the molecular weight of the chlorophyll molecules, a weight ratio of chlorophyll to carotenoid of approximately 4:1 to 6:1 would be expected. On the other hand, the carotenoid molecules are slender linear molecules, about 5 Å in diameter, and therefore more than one molecule could conveniently fit into the 15 Å × 15 Å hole formed by the chlorophyll tetrads. From symmetry one might expect as many as four molecules per hole, but this number would lead to very tight fitting, which is energetically improbable. One can therefore put a lower limit on the number of chlorophyll to carotenoid molecules of roughly 1:1 and a weight of 2:1.

In Table 8.3 we see that the mono- and di-galactosyl diglycerides account for the major lipids in the chloroplast. These lipids, because of their properties, can form a lipid or lipoprotein matrix for the chlorophyll monolayers. From spatial considerations the ratio of two galactosyl diglyceride molecules to one chlorophyll molecule could stabilize all the chlorophyll molecules in the monolayer. That is, there would be one phytol chain of chlorophyll for four *cis*-unsaturated acyl chains of galactosyl diglyceride (Rosenberg, 1967). Such a relationship fits with the molecular model for chlorophyll in the chloroplast membrane. There are of course other possible ways in which the chlorophyll molecules could be oriented in the chloroplast lamellae. If the porphyrin parts of the chlorophyll molecules lie at 0°, as depicted in Figure 8.7, their greatest cross-section would be available. If they are oriented at increasing angles to 45° (Fig. 8.7b), the cross-sectional area would be decreased to about 100 Å². Because the chlorophyll molecules in the chloroplast are in a dynamic state, they would orient themselves for maximum light absorption, so their greatest cross-section would be available for light capture.

The precise roles that chlorophyll, carotenoids, proteins, enzymes, and lipids play in the chloroplast membranes during photosynthesis is not com-

Table 8.3 The Composition of Spinach Chloroplast Quantasome

Chlorophylls		230
Chlorophyll a		⌈160
Chlorophyll b	230————	⌊ 70
Carotenoids		48
β-Carotene		14
Lipids		740
Phospholipids		⌈116
Sulpholipids	164————	⌊ 48
Digalactosyl diglyceride		⌈114
Monogalactosyl diglyceride	460————	⌊346
Nitrogen atoms as protein		9380

Source: Wolken, 1975, p. 79 and based on data of Park and Biggins, 1964, p. 1010.

pletely understood, but studies to unravel their function continue to be actively pursued.

ATTEMPTS TO DUPLICATE PHOTOSYNTHESIS

> He had been eight years upon a project for extracting
> sun-beams out of cucumbers, which were to be put into
> vials hermetically sealed, and let out to warm the air.
> Jonathan Swift, *Gulliver's Travels,* 1947, p. 203.

Ever since scientists began to probe how chloroplasts utilize solar energy and convert the light energy to chemical energy they have wanted to duplicate the process outside the living cell. Even though this duplication has not yet been accomplished with an efficiency comparable to living plant cells, various models have been and are being assembled that mimic the chloroplast system, converting light to chemical and electrical energy. There is no reason to believe that the photosynthetic process outside the living cell cannot be achieved in time.

One approach is to reassemble the chloroplast from the analysis of its composition and to recreate the molecular structure in which chlorophyll accessory pigments, enzymes, lipids, and proteins are assembled in the chloroplast membrane. The assembly must be arranged in such a way that light energy is captured by a photosensitive dye resulting in photochemical and photoelectric processes. From this process a gradient of chemical potential can be generated via photochemical oxidation and reduction reactions, similar to the photosynthetic electron transfer scheme (Fig. 8.4). Attempts to isolate photoactive fractions from the chloroplast and to separate the fractions that contain Photosystems I and II have been actively pursued for more than a decade and some success has been achieved.

Another approach that we have suggested and tried (Wolken, 1984; Brown and Wolken, 1979) is that of a liquid-crystalline system. This system can be assembled to perform some of the photometabolic processes associated with photosynthesis. Chlorophyll molecules will orient in liquid crystals (Journeaux and Viovy, 1978) and a chlorophyll-lipid-protein can be structured into a liquid crystalline system (Ke and Vernon, 1971; Wolken, 1967). Surfactants, when dispersed in water, will form concentric lamellar structures like that of the phospholipid bilayers. Of the nonionic surfactants, digitonin, a digitalis glycoside whose structure resembles cholesterol, is of interest. Digitonin (1-2%) in water forms micelles that have a strong attraction for many complex molecules, particularly lipids and naturally occurring pigments like carotenoids and chlorophyll. The role of the digitonin micelle is to react with one of the substrates while simultaneously attracting the other substrate to the same vicinity. This role parallels the behavior of an enzyme in bringing the reactants together. The digitonin micelles can be used then for biological assemblies because the interactions responsible for micelle stability are similar to those that stabilize bioaggregates. These supramolecular assemblies compartmentalize reacting molecules and have a pronounced catalytic effect on energy and electron transfer reactions by virtue of the potential gradients at the interface.

The extracted chlorophyll from chloroplasts in digitonin micelles is chloroplastin. Chloroplastin is birefringent when observed through crossed polarizers; hence, there is an alignment of the molecules in the digitonin micelle. If a drop of chloroplastin is evaporated rapidly from a surface, lamellae are formed. When the chloroplastin lamellar structure is scanned with a microspectrophotometer at 675 nm (the major absorption peak for chlorophyll), the chlorophyll is found to be concentrated and oriented within the lamellae and not in the interspaces (Wolken, 1975; Brown and Wolken, 1979).

Chloroplastin is photoactive; it can photoreduce a dye, evolve oxygen, and in the presence of the right cofactors, perform some of the primary steps of photosynthesis, such as turning inorganic phosphate to organic phosphate, ATP (Serebrovskaya, 1971; Wolken 1966a, 1967, 1975).

Attempts to duplicate photosynthesis outside of the living cell have used chlorophyll and other pigments together with a variety of cofactors, enzymes, and proteins. Calvin (1983) has reviewed the problem of trying to create an artificial photosynthetic system in a membrane for light capture, conversion, and storage. Chlorophyll is complexed with a specific protein in the chloro-

Table 8.4   The Chlorophyll-Protein Complex Spectra

Protein	Maximum Absorption Peak	Fluorescence
Gelatin	673	
Peptone	673	
Albumin (bovine)	675	
Egg white (avian)	673	+
Egg yolk (avian)	674	+
Horse serum	675	+
Globulin (bovine)	675	
+ Globulin (rabbit)	675	
II + III Globulin (rabbit)	675	
IV, 1 Globulin (rabbit)	673	
Globulin (ox)	668	

Source: Wolken, 1958, p. 94

plasts, and a number of semisynthetic chlorophyll-protein complexes have been prepared using a variety of different proteins. These chlorophyll-protein complexes were used to study the oriented properties of the chlorophylls and their electron energy transfer by following shifts in their spectral absorption peaks (Table 8.4). Many of these chlorophyll complexes show promise but their photosynthetic efficiency is very low. Thus, we are learning how to reproduce a system capable of photosynthesis outside of living plant cells.

REMARKS

To understand how photosynthesis works by photophysics (the direct transduction of solar energy to chemical energy) has been a great challenge to scientists over the past two centuries.

It has not yet been explained how the chlorophylls, carotenoids, lipids, enzymes, proteins, and other components of the chloroplast fit into the particular molecular arrangement for function in the two photosynthetic systems. However, by putting the bits of information together, answers may be found to the question of how the photoexcitation and photochemistry takes place within the chloroplast membranes to initiate the photosynthetic process.

In this brief review of photosynthesis it is clear that many fascinating but perplexing problems remain to be solved. How did the photosynthetic first system evolve? How do the physical and chemical reactions of these photosystems precisely function? And finally, when these are understood, how can we reproduce photosynthesis outside of the living cell?

9

Searching for Light

> How a nerve comes to be sensitive to light, hardly
> concerns us more than how life itself originated; but I
> may remark that, as some of the lowest organisms, in
> which nerves cannot be detected, are capable of
> perceiving light.
>
> Charles Darwin, *The Origin of Species,* 1859.

PHOTOTACTIC MOVEMENT

Living organisms, from bacteria to animals, search for light. They move or orient themselves to maximize the sun's light. For example, on any beach humans are continually aligning themselves in the direction of the sun for suntanning and warmth.

Early in the history of life, unicellular organisms evolved the means for movement and photoreception that enabled them to respond to light in search of an optimum environment—all in an effort to increase their chance for survival. The patterns of phototactic movement in response to light stimuli bear directly on the underlying mechanisms of more highly evolved plant and animal photosensory systems. Therefore, isolating the function of light in phototactic behavior is of considerable help in our understanding of how light affects the behavior of living organisms.

Phototactic movement can be separated into phototropism and photo-taxis. Phototropism is simply a positive or negative orientation of a part or of the whole of an organism, either to or away from a light stimulus. An example of positive phototropism is the bending and twisting of a plant leaf to present its surface to the light. The ability of a plant to orient in search of light increases its efficiency for photosynthesis, which in effect determines the plant's biochemistry and is a factor in the plant's growth.

More than a century ago, Julius von Sachs (1864) observed that the bending of plants toward light is stimulated primarily by blue light. It was then found that blue light induces phototropism in fungi, whereas in the phototropism of ferns and higher plants both blue light and red light are required; for mosses, only red light seems to be effective. Because light must first be absorbed to produce an effect, the plant organs that exhibit phototropism should contain blue-absorbing and red-absorbing pigments. So, what are these photoreceptor molecules for phototropism, and do they reside in photoreceptor structures?

Fungi are widespread in nature and most seek light—they bend and turn toward light. Fungi do not synthesize chlorophyll and are not capable of photosynthesis. Their nutrition for growth is heterotrophic, being either saproprytic or parasitic. Thus, we can isolate their phototropic behavior without the complexity of photosynthesis.

PHYCOMYCES, A MODEL PHOTOTROPIC CELL

It is of interest to review some experimental studies that we have pursued in search of the photoreceptor system for phototropism with the fungus, *Phycomyces blakesleeanus* (Wolken, 1972, 1975).

The sporangiophore of *Phycomyces* is a single aerial cell that matures in distinct stages designated I-IVb, and in the process grows to more than 10 cm in length and is about 0.1 mm in diameter (Fig. 9.1*a*). The growth stages are related to the elongation of the sporangiophore and to the development of the sporangium, the spore head. In the maturation processes, color changes are observed in both. The general appearance of the sporangiophore in stage IVb is that of a nearly transparent cylindrical filament supporting a spherical black sporangium. The growth time from stage I to stage IVb is dependent on light, temperature, humidity, and the growth medium. The sporangiophore is light-sensitive, exhibiting phototropism when it is illuminated unilaterally with light that it can then follow (Fig. 9.1*b*). It is positively phototropic to wavelengths from 300 nm to 510 nm, and negatively phototropic to wavelengths shorter than 300 nm.

Growth is stimulated by light, so light functions for the organism as a signal to alter its growth in space and time. For example, light-searching with growth is observed when a colony of *Phycomyces* is cultured in darkness until it reaches stage I, and then a short flash of light is given through a pinhole. When they reach stage IVb, all the sporangiophores are found to be

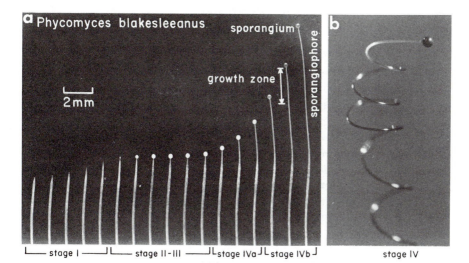

Figure 9.1 *Phycomyces blakesleeanus. (a)* Growth study over time; *(b)* demonstration of helical phototropism, stage IV. (Courtesy of Professor David Dennison, Dartmouth College, Hanover, New Hampshire.)

bent toward the pinhole. Therefore, in the absence of light, they continued to search for the direction of the light signal. In a sense it could be said that these *Phycomyces* exhibited "memory"—for in stage I they were photosensitive, and their photoreceptor system was capable of "remembering" the direction of the light signal hours later when their growth reached stage IVb.

Besides responding to light, the sporangiophore is also sensitive to gravity, to touch, and to the presence of nearby objects. Why is all this sensory behavior built into such a primitive organism? Because it can so keenly sense its environment, what can be learned from it about the development of more highly evolved plant and animal photosensory systems? As an organism, it possesses no obvious structure that resembles an eye or a nervous system. What are the photoreceptor molecules and receptor structures that permit the organism to detect environmental changes, behavior we usually associate with animals? In search of answers to the questions raised, mutants can be produced by physical and chemical means and strains selected that are (1) fully sensitive to light, (2) night-blind (sensitive to only high light intensity, as are the cones in the retina of vertebrate eyes), and (3) insensitive to light and considered blind. If one can establish here some genetic basis at the molecular level for the ability to see or for blindness, then a tool is available to investigate photosensitivity by the methods of molecular genetics. From investigation of these mutants, clues are being obtained on the biosynthesis of photoreceptor pigment molecules and photosensitivity.

The Photoreceptor Molecule

What is the photoreceptor molecule for phototropic movements? The photosensory region is the growth-zone of the sporangiophore. It extends from 0.1 mm to about 3 mm below the sporangium and occupies a surface area of about 6×10^{-3} cm^2. The sporangiophore is most photosensitive in states I and IVb. The action spectra for *Phycomyces* phototropism (Fig. 9.2b,c) shows absorption peaks around 280, 365–385, 420–425, and 445–485 nm (Delbrück and Shropshire, 1960; Curry and Thimann, 1961). So we infer that the photoreceptor molecule should have similar absorption peaks, and this is evidenced in Figure 9.2a. The absorption spectrum obtained by microspectrophotometry for stage IVb of the wild-type sporangiophore through the growth-zone is shown in Figure 9.3a. In scanning the sporangiophore from 0.1 mm to 2 mm below the sporangium, the absorption spectrum

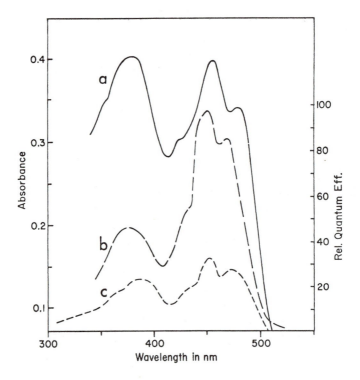

Figure 9.2 Absorption spectrum of an acetone fraction from *Phycomyces blakesleeanus (a)*, compared to action spectra for *Phycomyces* phototropism obtained by *(b)* Curry and Thimann, 1961, and *(c)* Delbrück and Shropshire, 1960.

gradually shifts to that of Figure 9.3*b*. The spectral peaks of Figure 9.3*a* resemble the absorption peaks of β-carotene at 430, 460, and 480 nm (Fig. 7.10), and correspond to those in the visible part of the action spectrum for *Phycomyces* phototropism. The absorption peaks in the ultraviolet near 280 and 370 nm, and in the visible around 450 nm, are similar to those of a flavin or flavoprotein molecule (Fig. 7.13). It is interesting to note that the absorption peaks of Figure 9.3*a,b* together give absorption peaks at 280, 370, 435, 460, and 485 nm, which correspond to the full action spectrum for phototropism (Fig. 9.2).

Phycomyces albino mutants are deficient in β-carotene, but are still phototropic, so their absorption spectra should be more informative. The absorption spectrum within the growth-zone of the mutant has absorption peaks near 230, 267, and 370 nm (Fig. 9.3*c*).

Neither the action spectra nor the absorption spectrum of the phototropic growth-zone of *Phycomyces* has been sufficiently informative to clearly

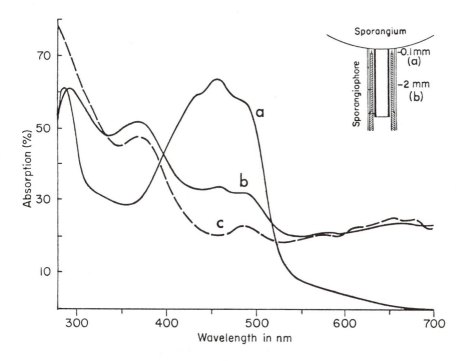

Figure 9.3 Microspectrophotometric scan showing absorption spectra in growth zone of *Phycomyces,* wild type, stage IVb, at *(a)* 0.1 mm to 2 mm and *(b)* from 2 mm and below the sporangiophore, compared to *(c)* albino mutant (C5) absorption spectrum in the growth zone.

identify the primary photoreceptor molecule. It is difficult to distinguish between carotenoids and flavoproteins because they have similar absorption spectra in the visible region, and even their fluorescence is inconclusive.

To determine whether flavins were present and in what concentration, the flavins were isolated from the sporangiophores. The total flavins of the wild type was 13×10^{12} and the albino mutant was 5×10^{12}. The flavins that were identified were riboflavin, lumiflavin, lumichrome, flavin adenine dinucleotide (FAD), and flavin mononucleotide (FMN). How reasonable are these findings about the sporangiophore? The microspectrophotometry of the growth-zone showed a shift in a spectral characteristic from a typical carotenoid spectrum (Fig. 9.3a) to a typical reduced flavin semiquinone spectrum (Fig. 9.4), suggesting that a flavin is indeed a photoreceptor molecule for phototropism (Wolken, 1969, 1972, 1975). Nevertheless, a carotenoid and still other pigment molecules may be involved in the photoprocess as well.

The Photoreceptor Structure

The *Phycomyces* sporangiophore is nearly transparent and could act as a cylinder lens. The evidence for a lens effect has been interpreted as indicating that the photoreceptor is located in or near the cell wall. Cohen and

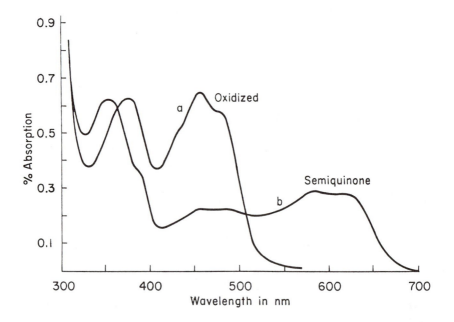

Figure 9.4 Absorption spectra of a flavoprotein extracted from *Azotobacter*, *(a)* oxidized and *(b)* semiquinone.

Delbrück (1959) showed that the primary effect of the light must be on some structure that moved relative to the cell wall. This finding necessitated a thorough microscopic search for a structure that moved in the growth-zone in response to light. Our observations with polarization microscopy revealed birefringent rodlike crystals (Fig. 9.5*a*, *b*) that were aligned near the vacuole in the growth-zone (Wolken, 1969). When the sporangiophore was permitted to dry, numerous needle crystals appeared throughout the sporangiophore (Fig. 9.5*c*). With fluorescent microscopy, both the rods and needle crystals were found to fluoresce. Another kind of crystal that was observed was

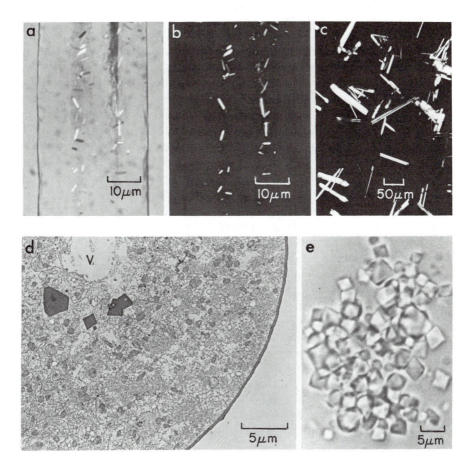

Figure 9.5 *(a)* Crystals in the light-growth zone of *Phycomyces*. *(b)* Same area as *(a)* in polarized light; *(c)* needle and rod crystals in polarized light; *(d)* electron micrograph of cross section through light-growth zone showing octahedral crystals; and *(e)* octahedral crystals isolated from organism.

octahedral in shape, and appeared to tumble in the sporangiophore (Fig. 9.5d,e), suggesting that these octahedral crystals could be the photoreceptor.

Further microscopic observations of the sporangiophores during growth indicated that octahedral crystals are found in all stages of sporangiophore development. The number of crystals per sporangiophore consistently increased with growth (stages II-III) and with the increase of sporangiophore length until early stage IV, then gradually decreased. In cultures grown in darkness, the number of octahedral crystals per fresh weight of sporangiophores was highest in stage I and lowest in stage IV. Also, the number of crystals in the sporangiophores was found to be greatly dependent upon the intensity of the light. Continuous illumination reduced the crystal content in sporangiophores at all stages of growth, resulting in practically a twofold difference between dark- and light-grown organisms.

It is interesting that there were more crystals in the wild type than in the β-carotene deficient mutants, and the smallest number was found in the night blind mutant, which is phototropic only at high light intensity. Keeping in mind the possibility that β-carotene functions well as a filter in the wavelength range of 400-500 nm, and that at high light intensity photodestruction of the crystals takes place, there should indeed be more crystals in the wild type, which contain a larger amount of β-carotene than the albino mutants.

Therefore, are the octahedral crystals, which are primarily found in the growth-zone, in fact the photoreceptors? The octahedral crystals from the sporangiophore can be isolated in a relatively pure state by sucrose density fractionation (Figs. 9.5e and 9.6). The absorption spectrum obtained from a single crystal by microspectrophotometry showed absorption in the ultraviolet around 280 nm, indicating a protein, and near 350 nm and 460 nm in the visible, indicating an oxidized flavoprotein (Ootaki and Wolken, 1973). Upon irradiation with light the spectrum shifted to that of a flavin semiquinone (Fig. 9.4). Chemical analysis of the crystals indicates that they are about 95% protein (of which 1% by weight is tryptophan) and that there are lipids, flavins, and carotenes associated with the crystals.

But, the question of whether the flavoprotein is the photoreceptor molecule remains. The absorption spectrum of a crystal from stage IVb shows absorption in the ultraviolet, around 280 nm, and near 360 nm. However, in stages I and II-III, the crystal spectra show absorption near 275-285, 355, and around 465 nm in the visible. These are, in fact, the same absorption peaks found in the phototropic action spectrum and by microspectrophotometry of the crystals in the sporangiophore (Figs. 9.2 and 9.3). These data suggest that a flavoprotein is the photoreceptor molecule.

Electrophysiology

Let us now investigate whether the *Phycomyces* sporangiophore is a photosensory receptor cell analogous to sensory cells of animals. If it is, then

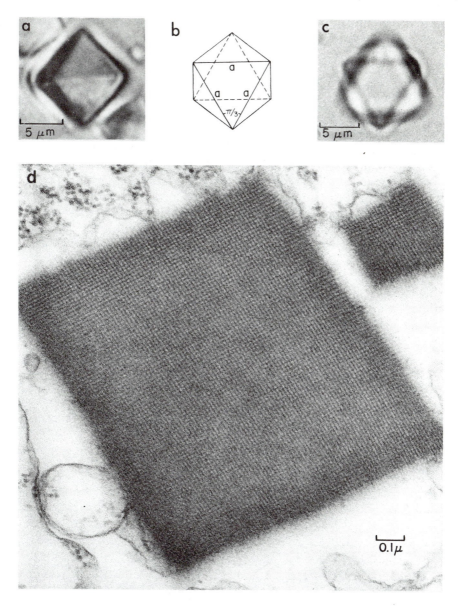

Figure 9.6   Octahedral crystals isolated from *Phycomyces*. Light microscopy of side view *(a), (b), (c),* showing equilateral triangular crystal faces; and *(d)* electron micrograph through the octahedral crystal showing lattice spacing in the crystal.

signals should be detected upon electrical stimulation and, as a photosensory cell, upon light excitation.

The electrical phenomena accompanying the light response of a photoreceptor may be divided into three types: (1) the early receptor potential (ERP), which is related to the photoreceptor pigment, (2) the receptor potential, which is a slow positive and/or negative potential, and (3) the receptor potential related to nerve spikes. These electrical signals have been studied extensively for photoreceptors and sensory cells of animals; in plants, early receptor potential responses have been found in the leaves of *Phaseolus vulgaris,* and *Mimosa* (Ebrey, 1967; Sibaoka, 1962; Arden et al., 1966).

The *Phycomyces* sporangiophore under continuous illumination decreases its sensitivity in response to a given flash of light. This loss of sensitivity, as with most photoreceptors, is some function of the intensity of incident light, is reversible, and indicates change in a photosensitive pigment that is bleached by light and is resynthesized in the dark.

Sporadic early receptor potential type response occurs upon illumination from a high intensity light source (60 watt/sec). The most prominent response for all growth stages studied was a graded receptor potential, a positive wave whose amplitude was from 2 mv to 10 mv and 2 sec to 10 sec in duration (Fig. 9.7). The amplitude correlated roughly with light intensity. At equal intensities the amplitude showed a larger response to a wavelength of 485 nm than to wavelengths of 420 or 385 nm. A receptor potential was obtained in response to 500 nm in stage I, but not in stage IV.

To see if the cell was electrically excitable, square wave pulses of 30 mv amplitude and 1 msec duration were applied to the sporangiophore. For

Stage	Light Source	Intensity	Response
IVb	microflash	60 watts/sec.	a —— 10mv. 1 sec.
IV	fluorescent lamp	250μ watts/cm^2	b —— 10mv. 2 sec.
I	monochromator 485 nm	150μ watts/cm^2	c —— 10mv. 2 sec.
I	monochromator 420 nm	150μ watts/cm^2	d —— 10mv. 2 sec.
I	microflash 420 nm	60 watts/sec.	e —— 10mv. 2 sec.

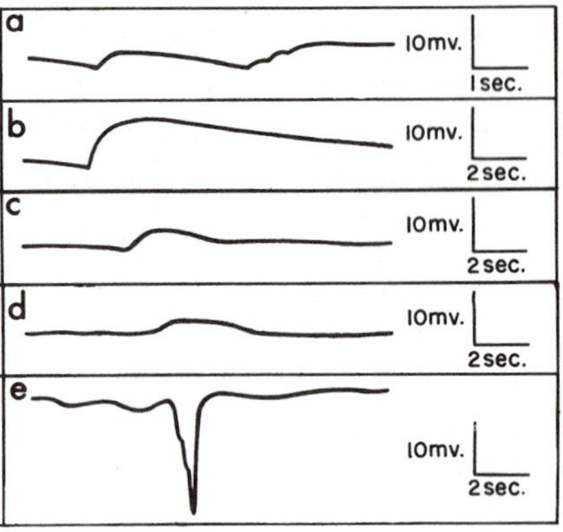

Figure 9.7   Electrical responses of *Phycomyces*.

stages I, IVa, and IVb, small amplitude, biphasic potentials were seen in response to the electrical stimulation, rarely exceeding 10 mv and occurring with an average latency of 1 min. As the sporangiophore matured through stage II to IVb, the complexity and type of the observed electrical response appeared to change. Although the receptor potential was seen for all stages studied, the sporadic early receptor potential type response and the spike were not.

These results indicate something about the pigment system of *Phycomyces,* because orientation of the photoreceptor pigment molecule is related to the early receptor potential. The latency and time course of the receptor potential are slow when compared to those of other photoreceptors (Tomita, 1970). The amplitude of the response is related to both wavelength and intensity (Fig. 9.2*b,c* and 9.7).

These electrophysiological results are of interest in considering whether the *Phycomyces* sporangiophore is a kind of photoneurosensory cell. If so, some sort of neurotransmitter molecule should be found, such as acetylcholine, the one we are most familiar with in more advanced photosensory cells (See Fig. 9.14).

Experiments were carried out to isolate the enzyme acetylcholenesterase, and our findings showed that there are 10^{-8} moles/gram, comparable to its concentration in brain tissue (10^{-6}/moles/gram). So we see that even in this primitive organism both a photoreceptor and the neurochemistry are found to be typical of more highly evolved specialized neurosensory cells (Mogus and Wolken, 1974).

PHOTOTAXIS

Let us turn from phototropism to phototaxis in search of photosensory cell mechanisms. Phototaxis is observed in free-moving organisms (bacteria, algae, protozoa and aquatic animals) when they move to or away from a light source. The organisms orient with respect to the direction of light by determining the intensity and wavelength of the light source.

The swimming patterns of microorganisms have fascinated microscopists for a long time. Antoni van Leeuwenhoek in 1674, using his simple microscope, described the swimming of algae and protozoa.

"These animalcules had diverse colors, some being whitish and transparent; others with green and very glittering little scales; others again were green . . . . And the motion of most of these animalcules in the water was so swift and so various, upwards, downwards, and round about, that 'twas wonderful to see . . . ." (Leeuwenhoek, 1674-1675, quoted by Dobell, 1958, pp. 110-111).

Such observations have continued to intrigue biologists, for from these observations have come interesting relationships among the various behavioral responses to light. Therefore, the analysis of phototaxis has broad implications toward our understanding of sensory mechanisms of more highly evolved animal cells.

Bacteria, algae, and protozoa are the simplest organisms for quantitative studies of the relationship between light stimuli and their behavioral responses. An ideal organism for such investigation in *Euglena gracilis* (Wolken, 1967, 1975). When *Euglena* are observed through a microscope, various kinds of motion can be distinguished: pulsation, sidewise rotation, and forward swimming. The sidewise rotation and forward swimming are controlled by the whipping of the flagellum at the anterior end of the organism. Light acts as a stimulus and *Euglena* respond by swimming toward the light source.

The Photoreceptor System

Euglena is structured so that it has a photoreceptor for light detection and an effector, the flagellum, so it can change the direction of its movements. The photoreceptor system consists of the eyespot, the paraflagellar body, and the flagellum (Fig. 9.8). From *Euglena's* reaction to light it was suggested some time ago that the photoreceptor pigment inside the eyespot was photosensitive and functioned similarly to the visual pigment of animal eyes (Englemann, 1882; Mast, 1911).

The *Euglena* eyespot is an orange-red body at the anterior end of the organism. It is an agglomeration of numerous pigment granules, each of which is about 0.1-0.3 mm in diameter. The cross-section of the eyespot area is about 6 mm^2. The granules of the eyespot are located just below the membrane of the reservoir, a chamber with smooth walls that follows the ridged gullet from where the flagellum originates (Fig. 9.9a). The eyespot granules are thought to function as a shading device and filter for the photoreceptor. A dense, homogenous body attached to the flagellum and facing the eyespot granules is the paraflagellar body, the organisms photoreceptor (Fig. 9.9a-d). The paraflagellar body, as observed with the electron microscope, is a crystalline structure.

How the light energy absorbed by the photoreceptor is transduced to chemical energy for flagellar activity in phototaxis is not precisely known.

The Search for the Photoreceptor Molecule

The degree to which *Euglena* responds to different intensities and wavelengths of light has been experimentally measured. As we have already seen, spectral sensitivity, or phototaxis action spectrum, corresponds to the absorption spectrum of the molecule or molecules responsible for the phototactic

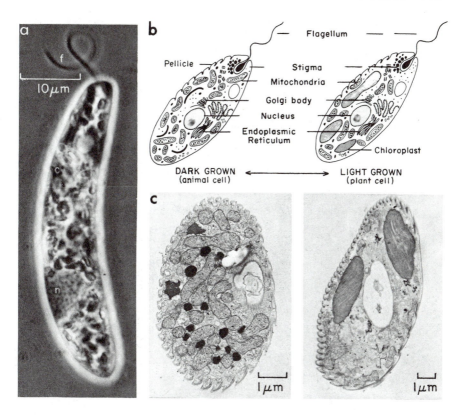

Figure 9.8 *Euglena gracilis (a),* cross-sections *(c)* of dark-grown and light-grown (electron micrographs) whose stuctures are schematized in *(b).* Refer to Wolken, 1971, p. 22 and Wolken, 1975, pp. 52–55.

behavior. So, in a sense we can communicate with the organism in that its speed and direction are controlled by the intensity and wavelength of light. The photobehavioral responses were separated into photokinesis and phototaxis for experimental purposes. Photokinesis is a change in the velocity of swimming upon illumination, without regard to orientation, whereas phototaxis is the orientation of the entire organism with reference to the light stimulus.

The photokinesis action spectrum for the rate of swimming (mean velocity in nm/sec versus wavelength at 6 watts/cm^2 intensity) has a major peak at 465 nm, and another near 630 nm (Wolken, 1975). The peak at 465 nm suggests a carotenoid (β-carotene), whereas the peak near 630 nm suggests chlorophyll or perhaps other molecules that participate in these photoprocesses.

When the action spectrum for phototaxis is compared to the action

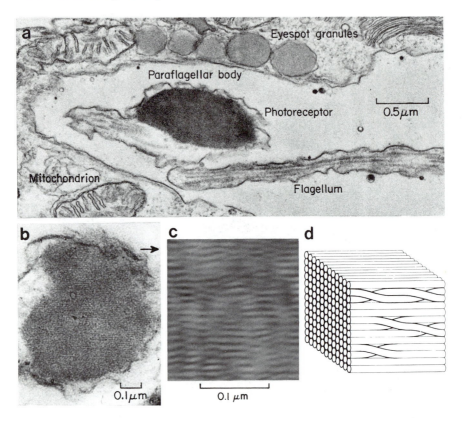

Figure 9.9   *Euglena gracilis. (a)* Eyespot area for phototaxis, eyespot granules, paraflagellar body, and flagellum. *(b)* Paraflagellar body, the photoreceptor. *(c)* Paraflagellar body, filtered image obtained by optically diffracted electron micrograph. *(d)* Schematic drawing of orientation of fibers (lamellae) in the paraflagellar body. *(From Wolken, 1977, p. 519.)*

spectrum for O_2 evolution, the peaks are almost identical (Fig. 9.10), suggesting that *Euglena* searches for light to promote more efficient photosynthesis, and that similar molecules participate in both phototaxis and photosynthesis.

The Photoreceptor Molecule

Absorption spectra of the *Euglena* eyespot area obtained with the microspectrophotometer show a broad absorption band with peaks around 430, 465, and 495 nm in the visible, and around 350 and 275 nm in the ultraviolet (Fig. 9.11*a*). In the heat-bleached (HB) mutant that lacks chloroplasts, and

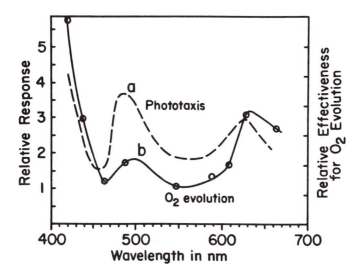

Figure 9.10 *(a) Euglena* phototaxis action spectrum; *(b)* compared to action spectrum for O_2 evolution in photosynthesis.

hence chlorophyll, the eyespot area spectra show similar peaks (Fig. 9.11*b*). Spectra closer to the paraflagellar body show absorption peaks at 440 and 490 nm and around 350 nm (Fig. 9.11*c*). This latter spectrum resembles that found for *Phycomyces* phototropism. When these light-grown *Euglena* are dark adapted for 1 hr and mounted on the cold stage (5°C) of the microspectrophotometer, and the same area is then illuminated with strong white light for 1-5 min, the absorption peak around 490 nm bleaches, accompanied by an increase in absorption at 440 nm. This spectral shift on light bleaching is similar to that of a flavoprotein semiquinone, from the reduced state to the oxidized state (Fig. 9.12). In addition, the spectrum for the *Euglena* paraflagellar photoreceptor resembles that found for the *Phycomyces* octahedral crystals (Wolken, 1971, 1975).

If a photoreceptor molecule is a flavin, then its identities and concentration should indicate whether it functions as the photoreceptor. The total flavins extracted from *Euglena* is of the order of 10^8 molecules per cell (Table 9.1), falling within the range for the number of molecules found for all photoreceptors.

This total is not too surprising, for flavins are known to be associated with many photoreceptor processes. For example, flavins are associated with the electron transport system of chloroplasts in algae, with chloroplast movement, with light-enhanced respiration in the alga *Chlorella,* and with phototaxis of the dinoflagellate *Gyrodinium dorsum* and of the insect *Drosophila* (Wolken, 1972, 1975).

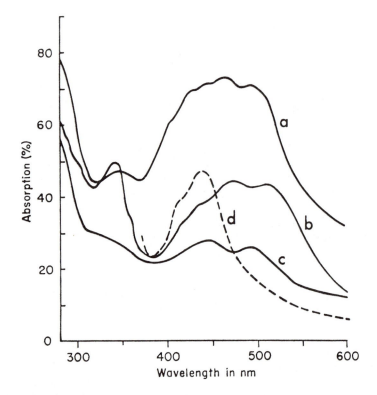

Figure 9.11 Absorption spectra obtained by microspectrophotometry: *(a)* eyespot, light-grown *Euglena; (b)* eyespot of heat-bleached (HB) *Euglena* mutant; *(c)* paraflagellar body, after 1 hr dark adaptation at 5°C; *(d)* spectrum of eyespot area after 5 min of white light.

There are also similarities in the photochemical process in the membranes of chloroplasts, in the oxidation electron-transport scheme between cytochromes and flavoproteins. Therefore, a photochemical mechanism can be conceived for *Euglena* in which a flavin and cytochrome *c* participate in the photoprocess. In such a photochemical system, a flavin can transfer reducing equivalents to cytochrome *c* as an intermediate carrier. The steps follow from a light-excited flavin to the reduced flavin that is oxidized by cytochrome *c;* the final step would be oxidation of cytochrome *c* by an O_2 molecule (Fig. 9.13).

There are other considerations with regard to the mechanisms of phototaxis. For the question remains, how is the light energy that is absorbed by the photoreceptor transduced to chemical energy that triggers flagellar motion? The chemistry of the algal flagellum indicates that it is composed of a myosin-

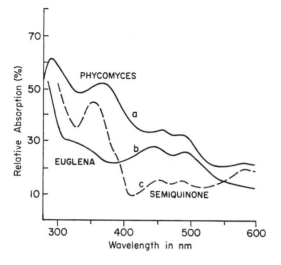

Figure 9.12 *(a)* Absorption spectrum of light growth zone, photoreceptor area of *Phycomyces,* compared to *(b)* spectrum near paraflagellar body of *Euglena. (c)* Compared to the spectral peaks of a flavin semiquinone.

Table 9.1 Flavins in Euglena

	Total Flavins $\mu g/cell \times 10^{-8}$
Wild-type, light-grown	4.52
Wild-type, dark-grown	5.06
Mutant, heat-bleached (HB)	4.60
Mutant, streptomycin-bleached (SM)	6.16
Mutant, Mg^{2+}-depleted	7.10

Source: Refer to Wolken, 1977, p. 521.

like contractile protein like that in muscle (Lewin, 1962; Witman et al., 1972; Witman, Carlson, and Rosenbaum, 1972). When large quantities of *Euglena* flagella are isolated and placed in 10^{-3} M solutions of ATP they display vigorous beating (Wolken, 1967, 1975). So we see that the flagellum possesses a mechano-chemical mechanism that converts the chemical energy of ATP into movement.

The *Euglena* eyespot region is structured much like a photoneurosensory cell, and if the mechanism of excitation is analogous to that of a neurosensory cell, it should then have the chemistry of nerve cells. Neurotransmitter

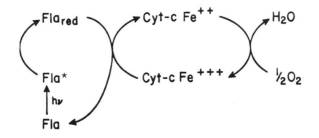

Figure 9.13   Possible photochemical mechanism. Arrows follow from light-excited flavin (Fla*) to a reduced flavin is oxidized by cytochrome *c;* the final step is oxidation of cytochrome *c.*

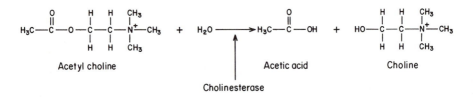

Figure 9.14   Acetylcholine and degradation by acetylcholine esterase to acetic acid and choline.

Table 9.2   Neurotransmitters in Protozoa

	Acetylcholine	Acetylcholinesterase	Catecholamines
Euglena gracilis	+	+	
Paramecium	+	+	+
Trypanosoma rhodesiense	+		
Tetrahymena geleii		+	
Tetrahymena pyriformis			+
Crithidia fasciculata			+
Noctiluca miliarus			+

Source: Based on data from T. L. Lentz, 1968, pp. 112-113.

molecules such as acetylcholine (Fig. 9.14) are in fact present in primitive organisms (Table 9.2), and there are 3.85×10^2 acetylcholine molecules per *Euglena* cell (Lentz, 1968; Wolken, 1971). These molecules are specifically associated with the chemistry of nerve excitation and the nervous system. It seems reasonable, because these substances are generally found in relatively primitive organisms, that the basic secretory capacity of these prenervous cells was modified during evolution, for the coupling of the excitatory

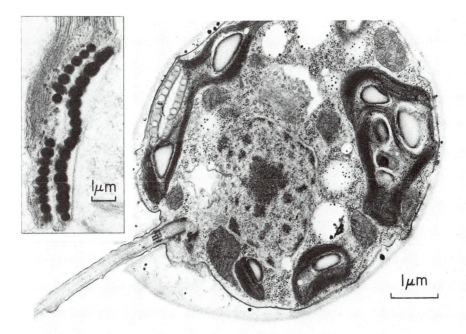

Figure 9.15 *Chlamydomonas,* cross section and enlargement of eyespot granules. Electron micrograph. (Courtesy of J. Jarvik, Carnegie-Mellon University, Pittsburgh, Pennsylvania.)

and conductile properties allows transmission to occur upon light and chemical stimulation.

REMARKS

The analysis of phototactic mechanisms has broad implications in our understanding of photosensory physiology.

In *Phycomyces* and *Euglena,* we found that the light detector is a crystal and the photoreceptor molecule is a flavoprotein. In the algal flagellate *Chlamydomonas* (Fig. 9.15), like *Euglena,* the photoreceptor system for phototaxis consists of an eyespot and flagellum. Unlike *Euglena*, the photoreceptor molecule in the eyespot is the carotenoid retinal, the chromophore of the visual pigment rhodopsin of animal eyes (Foster and Smyth, 1980; Foster et al., 1984). These experimental results confirm the early studies and speculations of Englemann (1882) and Mast (1911) that the photosensitive eyespot pigment was in fact the visual pigment. Even in the phototactic

Halobacterium halobium rhodopsin has been identified as the photoreceptor molecule (Oesterhelt and Stoeckenius, 1971). The finding of the visual pigment rhodopsin in primitive organisms will be further discussed in chapter 12.

So we see that the photoreceptor systems in light searching (i.e., the photoreceptor molecules for phototactic behavior in bacteria, fungi, algae, and protozoa) served other functions before they became adapted for photoneurosensory systems in animals. But much research needs to be done on phototactic behavior, from a molecular, structural, biochemical, genetic, and electrophysiological standpoint, all directed toward discovering the details of light reception and signal processing. The results will lead to a greater understanding of behavior in the more highly evolved photosensory systems of animals.

10

> When we reflect on these facts, here given much too
> briefly, with respect to the wide, diversified and
> graduated range of structure in the eyes of lower
> animals; and when we bear in mind how small the
> number of living forms must be in comparison with
> those which have become extinct, the difficulty ceases
> to be very great in believing that natural selection may
> have converted the simple apparatus of an optic nerve,
> coated with pigment and invested by transparent
> membrane, into an optical instrument. . .
> Charles Darwin, *The Origin of Species,* 1859

FROM SIMPLE EYES TO REFRACTING EYES

A major evolutionary step occurred in animals when the photoreceptor molecules became incorporated in the development of an eye. Animals with eyes were able to gain additional information about their world, thus greatly aiding their adaptation to life.

How did the eye evolve and how is it structured for vision? The development of the eye required an optical system, a lens to focus the light, and photosensitive receptor cells connected to a nerve to carry the imaged information to the brain.

Haldane (1966), in discussing the evolution of the eye, wrote

There are only four possible types of eye, if we define an eye as an organ in which light from one direction stimulates one nerve fiber. There is a bundle of tubes pointing in different directions and three types analogous to three well-known instruments, the pinhole camera, the ordinary camera with a lens and the reflecting telescope. A straightforward series of small steps leads through the pinhole type to that with a lens, and it is quite easy to understand how this could have been evolved several times.

The first eyes were regions of the outer skin that were sensitive to light. Pinhole-type eyes evolved from these. The pinhole eye is the simplest optical system for imaging. It uses the principle in which a small hole in the wall of an opaque chamber allows the passage of a very narrow beam of light from each point of an object, forming an inverted image on the opposite wall of the chamber (Fig. 10.1*a*). As an image-forming eye it is not very efficient, for only a small fraction of the light from an object can get to the photoreceptor surface. If the hole is made larger to increase the amount of light, image definition is lost, and if it is made smaller to improve the definition, diffraction becomes a problem. Still, this kind of eye has the advantage of simplicity, as no focusing whatever is required, and the size of the image is inversely proportional to the distance of the object.

For example, in the flatworms, the common planarian, *Platyhelminthes,* has two pinhole-type eyes, each containing pigment cells and photosensory cells. The pigment cells function to shade the sensory cells from light in all but one direction, enabling the animal to respond differentially to the direction of light; in this case, to move away from the light. The animal's tendency to avoid light is controlled by the balance of nervous impulses from the photosensory cells of the eye. Pinhole eyes are also found in the marine planarian *Convoluta roscoffensis* (Keeble, 1910; Wolken, 1971) and in *Nautilis*, a cephalopod mollusc.

Among the annelids, earthworms possess light-sensitive cells embedded in their body wall. Each cell contains a lens surrounded by a network of nerve fibers. In many leeches such cells are gathered into cups shielded by pigment cells. These cup-shaped clusters of light-sensitive cells lie under a lens, which is a specialized area of the skin. The complete structure of a lens and photosensory cell cluster is referred to as an *ocellus,* a simple eye (Fig. 10.2). Insects also possess from one to three ocelli. The median or dorsal ocellus of many insects consists of a layer of photoreceptor cells, a synaptic zone in which the axons of the photoreceptor cells come into contact with dendrites of the ocellar nerve fibers, and a nerve that leads from the eye to the brain (Chappell and Dowling, 1972).

Simple pinhole eyes of annelid polychaetes were described by Greef (1877), Andrews (1892), and later by Hesse (1899). The question arose whether such eyes are capable of forming an image. Hermans and Eakin (1974) used electron microscopy to investigate the structure of the eye of an alciopid, *Vanadis tangensis,* and Wald and Rayport (1977) then demonstrated that the alciopids *Torrea candida,* a surface worm, and *Vanadis,* found in the deep sea, possess image-resolving eyes. The annelid polychaete, *Odontosyllis enopla* of the family Sylladae in the Bermudas is of interest for it possesses a lunar periodicity and bioluminesces in its mating period. *Odontosyllis* is discussed further in chapter 15. The spectral response of its eyes to light (the electroretinogram, or ERG) showed that they can advanta-

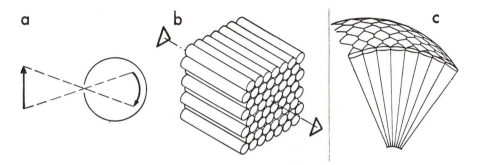

Figure 10.1 Optical imaging systems. Schematics of *(a)* pinhole, *(b)* parallel tubes, and *(c)* tubes arranged as in a compound eye. Refer to Figure 10.5.

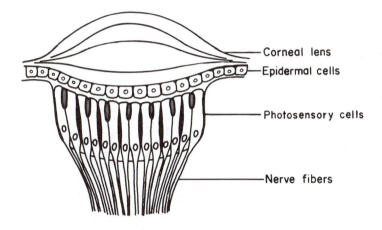

Figure 10.2 Photosensory cell. Schematic of an *ocellus*.

geously detect the bioluminescent emission spectral peak (Wilkins and Wolken, 1981). The fact that the worm can detect this bioluminescent light raised questions as to how the eye is structured for vision.

Odontosyllis has four eyes arranged so that two pairs are located adjacent to each other on the dorsal surface of the head (Fig. 10.3*a–d*). The eyes are on protruding lobes that can move. The eyes of the males are larger than the eyes of the females. The front of the eyes are almost completely covered by the worm's cuticle, and each eye has a rounded exterior with a relatively small opening and a lens located behind the opening. The lens is a spheroidal, gelatinous body lying in the cavity formed by the pigmented cup. Electron microscopy of the lens indicated that within the lens area there are

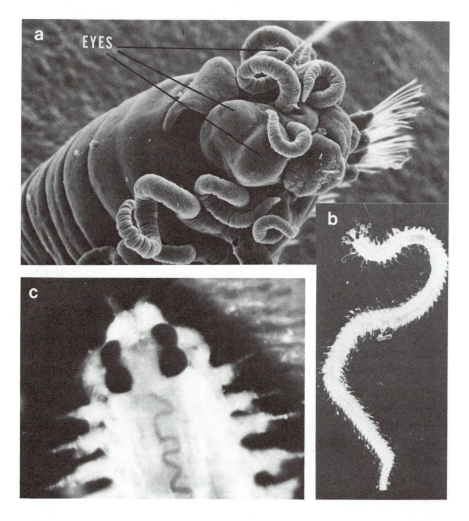

Figure 10.3 *Odontosyllis enopla* Verril. *(a)* Scanning electron micrograph showing head and eyes; *(b)* live female (2 mm long, 1.5 mm in diameter); *(c)* and *(c₁)* flattened area of the head showing the two eyes on each side of the head; *(d)* cross section through eyes; *(e)* enlargement of eye; *(f)* section through photoreceptor area, lens, photoreceptors and pigment granules; *(g)* enlargement of lens area; *(h)* microtubule in lens area.

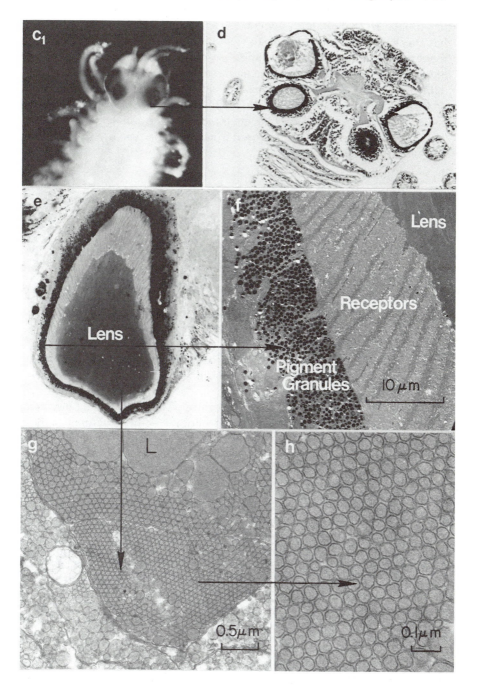

long tubular rods about 60 Å in diameter arranged in uniform linear arrays that are aligned with the optical axis (Fig. 10.3*e–h*), suggesting that they are fiber optic bundles. A fiber optic bundle system would function to detect the direction of the bioluminescent light and would maximize the light-collecting ability of the eye, for successful mating depends upon the detection and location of the bioluminescent light (Wolken and Florida, 1984). These studies of the *Odontosyllis* eye indicate that among the annelid polychaetes an evolutionary development occurred from a pinhole-type eye to a simple image-forming camera eye.

COMPOUND EYES

A second type of optical system is one in which an image is formed by a bundle of tubes that are separated by opaque walls. Only light falling on a particular tube in the direction of its axis can proceed to the end of the tube where it forms an upright image of the object. The image formed is the same size as the object, regardless of distance (Fig. 10.1*b*). Therefore, only those objects equal to or smaller than the optical device can be imaged completely, and no impression of distance by perspective is conveyed because the image size remains unchanged with object distance. However, if the tubes are disposed around the surface of a sphere or sphere segment with their axes pointing toward the center, as in image-forming compound eyes, then many of these disadvantages are avoided.

Compound eyes are restricted to the arthropods, which include insects, arachnids, and crustacea. The optics and photoreceptor structures of the compound eye are of considerable interest in elucidating the visual apparatus. The arthropod compound eye is very efficient for short-range vision and for detection in the total visual field. Many arthropods orient to the plane of vibration of polarized light and can discriminate colors.

Image-forming compound eyes evolved in the Cambrian era some 500 million years ago, as in the eyes of the trilobite fossil *Phacops* (Fig. 10.4). A surprising finding was that their lenses were calcite (glass) and that the eye had fiber optic elements directing the light to the photoreceptors (Stuermer, 1970; Towe, 1973).

Insects

The insect compound eye consists of eye facets *(ommatidia)* that vary from only a few in certain species of ants to several thousands, as in the dragonfly. Each ommatidium has a corneal lens (L_1), crystalline cone (L_2), and retinula cells that form rhabdoms (the retina), where the visual process is initiated. Each retinula cell has a differentiated specialized structure, the rhabdo-

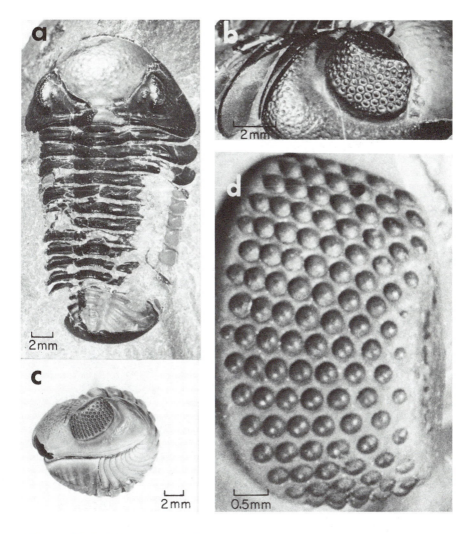

Figure 10.4   Trilobite fossil, *Phacops rana (a), (b), (c), (d)* compound eye structure.

mere, which is analogous in function to the retinal photoreceptors of vertebrate eyes.

Exner (1891) described two anatomically distinct types of insect compound eye structures, the apposition and superposition eyes. In Exner's model, apposition eyes are those in which the rhabdomeres forming the rhabdom lie directly beneath or against the crystalline cone, and an inverted image is formed at the level of the rhabdom (Fig. 10.5a). Each ommatidium is

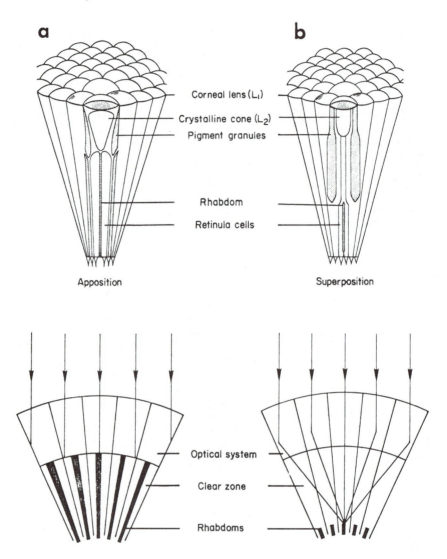

Figure 10.5 Compound eye structure, showing eye facets called ommatidia. (a) Apposition-type and (b) superposition-type compound eyes and their optics.

entirely sheathed by a double layer of pigment cells. Therefore, only light striking the lens within 10° of the perpendicular reaches the rhabdomeres. Light striking the lens at a more oblique angle is either reflected by the lens or absorbed by only the rhabdomeres of that ommatidium. This type of compound eye structure was believed to be characteristic of diurnal insects.

Exner's concept of a superposition eye was one in which the receptor cells reside a certain distance away from the crystalline cone (Fig. 10.5*b*). The term *superposition eye* has been loosely applied to any compound eye in which, during the dark-adapted state, the tips of the crystalline cones are separated from the rhabdom layer by an apparently clear space. The superposition eye was believed to be characteristic of nocturnal species because its mechanism was thought to be important for the requisite increase in light-gathering power. But, superposition eyes have been found in both diurnal and nocturnal species. Exner also suggested that the crystalline cone of superposition eyes had lens cylinder properties, the greatest index of refraction being at the axis of the lens cylinder, with concentric rings of decreasing indices of refraction as one proceeded to the periphery of the crystalline cone; this arrangement would be a graded index of refraction lens (GRIN lens). Such a system would allow for an erect image on the receptor layer, and would increase the light-gathering power, as the light from a point source entering the lenses of several ommatidia could be focused at a single spot on the rhabdom.

The number of rhabdomeres that form the rhabdom varies; in the open-type rhabdom there are from five to seven or more including an asymmetric rhabdomere lying in the same plane (Fig. 10.6). In the fused or closed-type rhabdom there are from four to eight rhabdomeres forming a symmetrical arrangement, whereas its asymmetric rhabdomere lies in another plane (Figs. 10.7 and 10.8). The closed-type rhabdom is most likely an efficiency mechanism used for light capture by nocturnal insects. In general, nocturnal insects such as cockroaches, fireflies, and moths have rhabdoms that occupy a large cross-sectional area, while daylight-active insects have relatively smaller cross-sectional area for their rhabdoms (Wolken, 1971).

The fine structure of all rhabdomeres is that of packed microtubules, from about 300 Å to 500 Å in diameter, whose double-walled membranes are from 50 Å to 100 Å in thickness. In the rhabdom, the microtubules of two adjacent rhabdomeres are oriented perpendicular to each other, whereas in the opposite pair of rhabdomeres, the microtubules are parallel (Figs. 10.5-10.8).

In certain insects, tracheoles surround the rhabdoms. They are seen in the June beetle (Fig. 10.9). The tracheoles surrounding the rhabdoms of moths are responsible for their eye glow, and in butterflies this glow appears red (Ribi, 1981). The ridges of the tracheoles function as a quarter-wave interference filter (Miller and Bernard, 1968) because the tracheoles are formed in a twisted lamellar pattern, like a screw, so that the thickness and

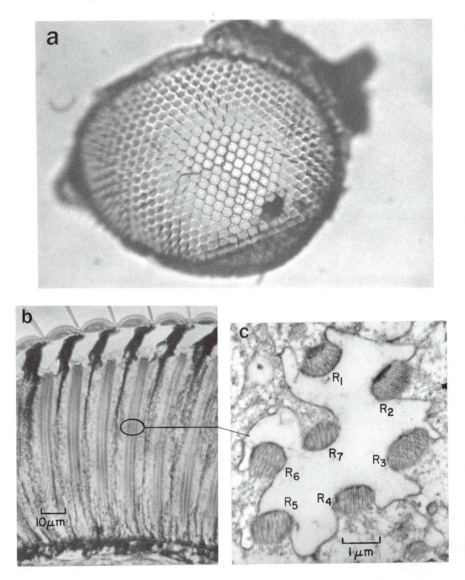

Figure 10.6 *Drosophila melanogaster, (a)* compound eye. *(b)* Longitudinal section through several ommatidia showing the corneal lens, crystalline cone, rhabdom, and pigment sheath; *(c)* cross section through the rhabdom illustrating the orientation and structure of the rhabdomeres (R_1–R_7).

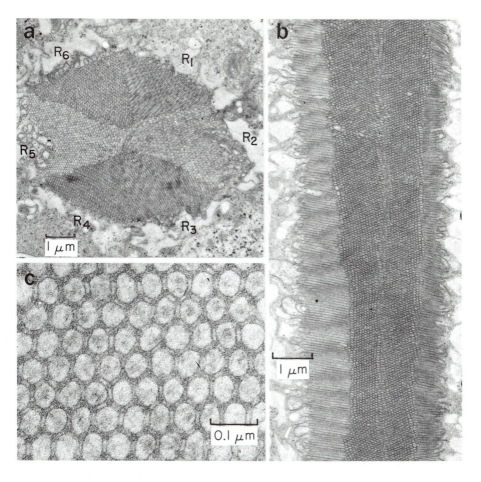

Figure 10.7 The carpenter ant, *Camponotus herculenus pennsylvanicus.(a)*
Cross section through the rhabdom showing rhabdomeres (R_1–R_6).*(b)* Longitudinal section of the rhabdom. *(c)* Higher resolution of a cross section
through many rhabdomere microtubules, electron micrographs.

spacing of the lamellae correlate with the reflected colors (Brown and
Wolken, 1979).

An interesting structural feature in certain insects is that the crystalline
cone and the rhabdom are connected by a fiber optic crystalline thread that
functions as a light guide. We find this thread in the firefly *Photuris* (Fig.
10.10), and the skipper butterfly, *Epargyreus clarus.* Only the light contained
in the crystalline thread is effective for stimulating the retinula cells (Horridge,
1975). Another feature of this fiber optic structure occurs in the worker

honeybee where the closed rhabdom and the surrounding zone act together as a waveguide.

Crustacea

Among the crustacea are found a variety of animals from the lobster and crabs, which are relative giants, to the tiny water flea only a few millimeters long. Like the insects, most species possess compound eyes. It is illuminating

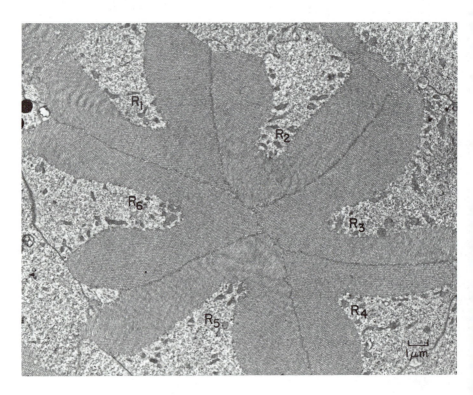

Figure 10.8 The firefly *(Photuris pennsylvanica)* rhabdom; cross section through the rhabdom showing geometric arrangement of its rhabdomeres R_1–R_6. *(From Wolken, 1971, p. 62.)*

Figure 10.9 (at right) June beetle (Scarab, *Phyllophaga*) cross section through compound eye in photoreceptor area, showing rhabdoms surrounded by tracheoles. *(From Wolken, 1971, republished in Wolken and Brown, 1979, p. 129.)*

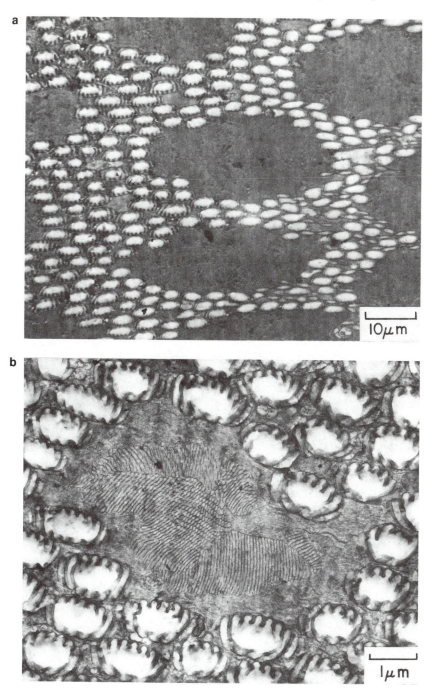

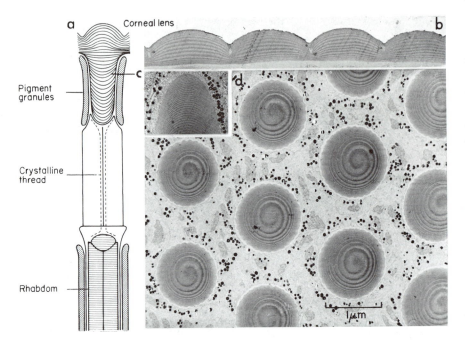

Figure 10.10   Corneal lens structure of the firefly *(Photuris pennsylvanica)* compound eye. Model of the ommatidium *(a)* in the compound eye. *(b)*, *(c)*, and *(d)* Electron micrographs of corneal lens structure. *(After Wolken, 1971, pp. 59, 60 and Wolken, 1975.)*

to describe several freshwater and marine crustacea that exhibit some general as well as special features in their eye structure.

The freshwater *Daphnia,* a water flea, possesses two eyes, a compound eye, and a simple nauplius eye (Wolken, 1971). The compound eye of *Daphnia pulex* and *Daphnia magna* consists of about 22 ommatidia enclosed in a capsule. The eye is in continuous oscillatory motion. The corneal lens, as seen in the insect ommatidium, and the distal pigment cells are not usually found. The ommatidium has a crystalline cone and elongated retinula cells that are surrounded by pigment granules. There are eight retinula cells that give rise to seven rhabdomeres to form the closed-type rhabdom (Röhlich and Törö, 1965).

The rhabdomeres that form the rhabdom consist of microtubules about 500 Å in diameter, as observed for the insect rhabdomeres. The microtubules within each rhabdomere are precisely arranged with their longitudinal axes regularly aligned in a given direction for one set of rhabdomeres, and in the perpendicular direction for alternate layers. This type of structure was observed in other crustacea, such as the land crab *Cardisoma* and the

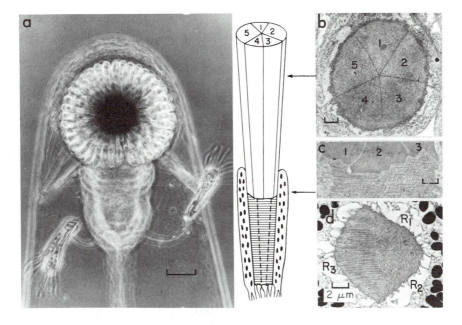

Figure 10.11 The waterflea, *Leptodora kindtii*. *(a)* The eye; the animal's cuticle acts as a common lens. *(b)* Cross section of crystalline cone, composed of five wedge-shaped segments (1–5). *(c)* Longitudinal section showing the connection of the crystalline cone with the rhabdom. *(d)* Cross section of the rhabdom indicating rhabdomeres R_1–R_3.

swimming crab *Callinectes* (Eguchi and Waterman, 1966). Such an arrangement of the rhabdomeres with their microtubules at right angles could be a structural basis for the ability of crustacea to analyze the direction of polarized light (Waterman, Fernandez, and Goldsmith, 1969).

Another freshwater crustacean is the carnivorous water flea, *Leptodora kindtii,* which is relatively large (18 mm in length) in comparison with *Daphnia.* The animal is almost completely transparent with one median spherical compound eye (Fig. 10.11). The entire eye of *Leptodora kindtii* is contained within the transparent exoskeleton at the anterior end of the organism. It is the exoskeleton itself that functions as the corneal lens. The eye is free to move and can rotate 10° in either direction (Wolken, 1975). A small area behind the eye with connecting neurons accommodates the optic process leading to the brain (Scharrer, 1964). *Leptodora* is believed to have better vision than most crustacea, for it can rapidly capture copepods as large and as fast as *Cyclops.*

The *Leptodora* compound eye is composed of approximately 500 ommatidia radially arranged (Fig. 10.11*a*). The ommatidia are large conical structures, measuring about 180 μm in length and narrowing from 30 μm in diameter at the outer portion to 2 μm at the base. A schematic ommatidium is illustrated and a cross-section through the crystalline cone and rhabdom is shown in Figure 10.11*b-d*. The crystalline cone constitutes about two thirds of the ommatidial length. The crystalline cone is composed of five equal wedge-shaped segments, formed from five crystalline cone cells that concentrate the light into a narrow beam.

The rhabdoms are affixed directly to the ends of the crystalline cones. There are four radially arranged retinula cells that form the rhabdom. Cross-sections through numerous ommatidia in the rhabdom area reveal that one of the rhabdomeres (R_3) is large in comparison with the other two and appears to be two rhabdomeres that have fused (Wolken and Gallik, 1965). The rhabdomere fine structure is that of tightly packed microtubules that average about 500 Å in outside diameter, and whose double-membraned wall is about 100 Å in thickness, like that found in the insect rhabdomeres.

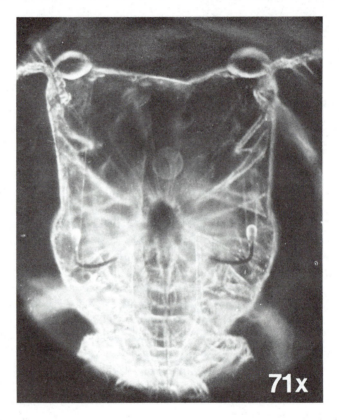

Figure 10.12   *Copilia quadrata,* Mediterranean, dark field photomicrograph. (Courtesy of Dr. Neville Moray, University of Toronto, Toronto, Canada.)

Among the crustacea is the relatively rare marine copepod, *Copilia* (Fig. 10.12), whose unique eye structure and scanning visual system has been of considerable interest ever since it was first found in the Bay of Naples and described by Grenacher (1879, 1886). Exner (1891), working in Naples, redis-covered *Copilia* and was fascinated by its strange pair of eyes, which he likened to a telescope. *Copilia quadrata* (about 1mm wide and 3 mm long) is found in the Mediterranean, and *Copilia mirabilis* in the Caribbean. Only the fe-male of the species possesses these wonderful scanning eyes, for the male is blind.

The *Copilia* eye structure has more recently been described by Vaissière (1961a, 1961b), Gregory (1966, 1967), and Wolken (1971, 1975). The eye re-sembles an ommatidium of the compound eye with a corneal lens (bicon-vex anterior lens, L_1) and, at some distance away, the crystalline cone (posterior lens, L_2). Attached to the crystalline cone are the retinula cells that give rise to the rhabdomeres that form the rhabdom, lying in an L-shaped, orange-colored stem that is the only pigmented part of the body (Fig. 10.13). It is this stem that oscillates back and forth in a saw-toothed pattern, varying from about 1 scan per 2 seconds to 5 scans per second (Gregory, Moray, and Ross, 1964; Moray, 1972). The stems from both eyes move synchronously and rapidly toward each other, then separate slowly. Gregory (1966) has likened such scanning to a television camera: "It seems that the pattern of

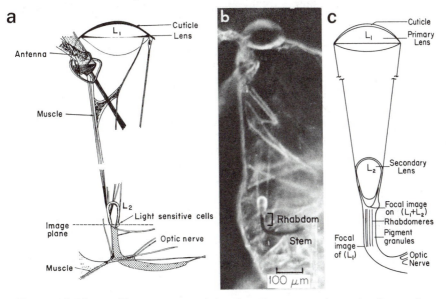

Figure 10.13 *(a)* The structure of the *Copilia* eye, as drawn by Grenacher (1879). *(b) Copilia quadrata,* dark field photomicrograph of the eye. *(c)* The optical system, showing the positions of the corneal anterior lens (L_1) and the crystalline cone, posterior lens (L_2); focal points of the corneal lens L_1; focal point of L_1 and L_2 in the total optical system. *(From Wolken, 1975, p. 159.)*

dark and light of the image is not given simultaneously by many receptors, as in other eyes, but in a time-series down the optic nerve, as in the single channel of a television camera (p. 201)."

It was expected that because *Copilia* lives at a depth (200–300 m) where the light level is near that of moonlight, it would have closed-type rhabdom with the significantly greater effective cross section necessary for better light gathering efficiency. But the *Copilia* rhabdom is the open-type, common to most insects that navigate at high light levels.

The *Copilia* eye is analogous to the superposition-type ommatidium (Fig. 10.5) in which the crystalline cone, or in this case the secondary lens, lies at some distance from the corneal lens (Fig. 10.13). In addition, the secondary lens forms a convex interface with a fluid of lower refractive index. The concentration of this material varies across its diameter, the greatest concentration being in the center, indicating that it has a graded index of refraction.

The *Copilia* eye with its corneal lens, L_1, and secondary lens, L_2, is a two-lens optical system in which the lens L_2 is positioned a short distance in front of the rhabdom and focuses the image on the rhabdom. In the design of such an optical system (Fig. 10.14) the corneal lens (L_1) forms an image at I, which is intercepted by the posterior lens (L_2) and is imaged as I_2. The nodal point of the combination *(N)* is found by drawing a line parallel to the original ray and passing through the final image; this line behaves like a ray passing through a single lens at N. The effective focal length *(EFL)* is the distance from N to I_2 and is obtained from the equation:

$$EFL = \frac{f_1 \times f_2}{f_1 + f_2 - s}$$

where f_1 and f_2 are the focal lengths (in water) of the anterior lens (L_1) and posterior lens (L_2), and s is the separation of their centers. From our data on

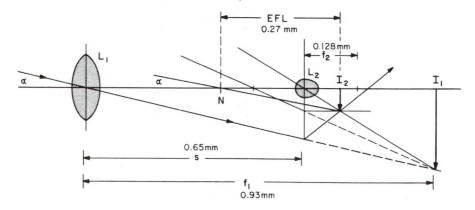

Figure 10.14 Optical system of the *Copilia* eye.

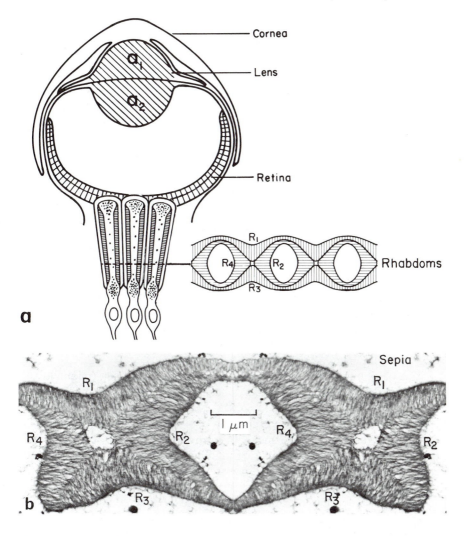

Figure 10.15 Mollusc cephalopod eye, as in the *Octopus*. *(a)* Note the structure of the retina. The lens is constructed of two halves (a_1–a_2) and the retina is formed of rhabdoms. Each rhabdom has four rhabdomeres (R_1–R_4). *(b)* Electron micrograph of the rhabdoms of *Sepia*.

the *Copilia* eye, $f_1 = 0.93$ mm, $f_2 = 0.128$ mm, and $s = 0.65$ mm. The *EFL* is then 0.27 mm. This value brings the *f*-number down to 1.6, comparable to that of a fish lens.

The importance of this optical system is that it greatly facilitates scanning and provides for a high-aperture and high-resolution optical device (Land, 1980). *Copilia* has adapted to very low light levels by evolving a remarkably advanced optical system that maximizes the collection of very diffuse light in its environment.

Molluscs

Following the insects and crustacea, the molluscs are by far the most numerous of the invertebrates. Included among the molluscs are clams, oysters, nautiluses, squids, and octopi. Almost every kind of imaging eye is found among them, from the simple pinhole eye of the nautilus to the refracting-type eye. The octopus attracts our attention, for it is a large muscular animal that can grow to more than 10 feet in length. The eyes are usually large compared to the animal's body size. The brain behind the eye is one of the most highly developed of any of the invertebrates and its behavioral responses are primarily visual ones. The octopus brain is divided into 14 main lobes that govern different sets of functions, and the optic lobes are the largest of all (Young, 1964).

The octopus uses both monocular vision and binocular vision. In the monocular mode, the eyes face in opposite directions with their long axes roughly parallel. In binocular mode the eyes are switched slightly forward relative to the body. So the octopus has a complete 360° field of vision. The octopus can easily distinguish one shape from another, and can achieve chameleonlike color matching with its surroundings (Packard and Sanders, 1969).

In physical organization the eye resembles the refracting-type eyes of vertebrates, and the lens is formed out of two halves joined together (Fig. 10.15). The photoreceptors are directly exposed to the incident light and the image on the retina is not inverted as in the vertebrate eye. The retina is like that of a compound eye, and retinal cells form rhabdoms, as we have described and illustrated for arthropod compound eyes.

POLARIZED LIGHT ANALYSIS

Many animals with compound eyes are able to analyze the plane of vibration of polarized light. The detection of the direction of this electric vector serves as a compass for navigation because the direction of polarization indicates the relative position of the sun with respect to the horizon. Sir John Lubbock (1882), an English banker, observed the behavior of ants and wondered how they found their way to and from their nests while foraging for food. His observations led him to conclude that ants used the sun as a compass for orientation and navigation. Although this explanation seemed to be reasonable, it was soon questioned, for how could insects find their way when the sun was hidden from view?

The question was not really answered until von Frisch's (1949) behavioral studies of the bee suggested that, in addition to the direction of a point source (the sun), insects could utilize polarization information from a patch of blue sky. To test his hypothesis, von Frisch (1950, 1967) constructed a model for the bee eye using eight triangular polarizing elements, each transmitting a quantity of light proportional to the degree of polarization. In his model, opposite pairs of rhabdomeres had their polarizers in parallel orientation (Fig. 10.16). As we have already seen, electron microscopy of insect, crustacean, and mollusc rhabdomeres all show this geometric arrangement of perpendicular and parallel microtubules, supporting von Frisch's model.

Assuming that the rhabdom is the analyzer of polarized light, then a creature's ability to detect polarization is dependent on the orientation of the visual pigment in the rhabdomeres. If the major axis of the visual pigment lies parallel to the tubule direction, and hence perpendicular to the normal incident illumination, perhaps an explanation for the rhabdom's action as a polarized light analyzer could be provided.

 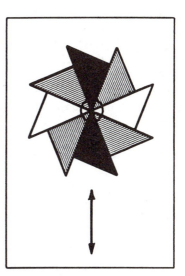

Figure 10.16   Model of the rhabdom with its rhabdomeres and how they can act as an analyzer for polarized light, in open and fused rhabdoms. *(Refer to von Frisch, 1950.)*

Houseflies and fireflies appear to orient to polarized light only within a narrow range of intensities, suggesting that there are two photoreceptors with different absorption vectors. If these two receptors are oriented at 90° with respect to each other, and if the long axis of the visual pigment molecule is oriented in a single plane but free to vibrate in that plane, the receptors will absorb polarized light in a ratio of two to one when the electric vector is parallel to the absorption vector of one of the photoreceptors.

Ants *(Tapinoma sessile, Solenopsis saevissimae)*, houseflies *(Musca domestica)*, fireflies *(Photurus pennsylvanicus)*, and Japanese beetles *(Popillia japonica)* all were found to orient to the plane of polarized light under conditions that prevented the use of background reflections as cues for orientation. These insects exhibited a compass reaction (the source of stimulation served as a fixed point by which an oriented path is maintained ±45° with respect to the plane of polarization of the incident light beam). All oriented at 0° and 90° (Marak and Wolken, 1965).

A white background reflects normally incident polarized light in a circular pattern, but a black background reflects elliptically. With the white background there was more orientation at 45° and less at 0° and 90°. This response difference is consistent with the hypothesis that there are more cues for simple phototactic response on the black background that reflects more light perpendicular to, rather than parallel to, the plane of polarization.

The reflectance pattern of nonpolarized light is circular, while the reflectance pattern of polarized light is elliptical, with the long axis perpendicular to the plane of polarization. In normal daylight, the long axis of the ellipse will point to the solar azimuth, indicating that insect navigation can be accomplished extraocularly.

A number of other hypotheses concerning the mechanism of directional analysis of polarized light have been suggested. One is that the direction of the electric vector of a beam of plane polarized light could be perceived through simple intensity discriminations because the direction of vibration is resolved into intensity gradients by reflection from the background. In fact, reflection patterns from the environment do resolve polarized light into patterns of graded intensity.

A model that is dependent upon reflection-refraction at the air-corneal lens interface brings us to examine the structure of the corneal lens. The corneal lens of the firefly *(Photuris pennsylvanica)* and the June beetle *(Scarab, Phyllophaga)* shows that it is formed of a laminated series of paraboloids, each lamella appearing to be rotated as pictured in Figures 10.10 and 10.17a. In transverse section, lamellae form a single or double spiral. Similar spiral structures were observed for the carpenter ant *(Camponotus herculenus)*, the fruitfly *(Drosophila melanogaster)*, and the housefly *(Musca domestica)*. A structural model for the corneal lens of these insects is illustrated in Figure 10.17b. The corneal lens' helical structure has the

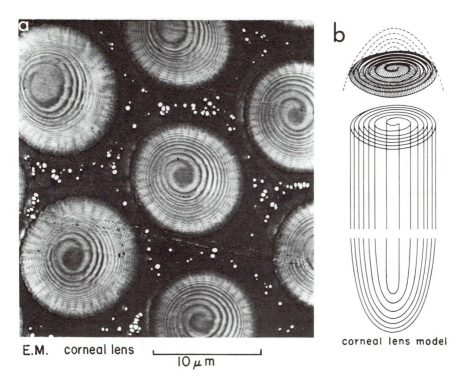

E.M.   corneal lens

|⊢————————————————⊣|
10 μm

corneal lens model

Figure 10.17 *(a)* Corneal lens structure of the June beetle (Scarab, *Phyllophaga*). *(b)* A schematized structural model of the corneal lens.

correct spacing and pitch to function as a polarizer of visible light, whereas the rhabdom, because of the way it is structured, (Figs. 10.6 and 10.8) would serve as the analyzer for the polarized light. Together they could function as a navigational device.

REMARKS

Only a few examples of invertebrates were selected to illustrate the diversity of eye structures in worms, insects, crustacea, and molluscs. Various optical systems for light collecting, focusing, and imaging are found in these animals for adaption to their environment, including pinhole, camera, compound, telescopic, and refracting-type eyes. The design of their optics is very sophisticated, utilizing fiber optics and aspherical nonimaging light collectors. It is of interest to point out that the compound eye corneal lenses are hexagonally packed structures, as in all insects (Figs. 10.1c and 10.18a),

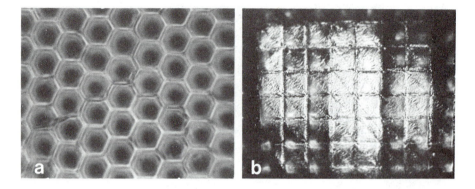

Figure 10.18 The corneal lens of *(a) Drosophila,* and *(b)* the lobster eye. Note the difference in packing of hexagons and squares.

while among the crustacea (shrimp, crayfish and lobster) they are squares. For example, the surface of the lobster eye is radially oriented squares (Fig. 10.18*b*). According to Vogt (1977) and Land (1980) light entering the crystalline cone, rod pyramid, is reflected from one or more of its sides that is focused on the photoreceptors, the rhabdom. Accordingly, it is highly probable that these square corneal lenses function as mirrors and the eye behaves optically as a reflecting superposition eye (Fig. 10.5).

The optics and photoreceptor structures of these invertebrate eyes suggest a number of interesting relationships in the evolution of the vertebrate eye and how we see.

Cajal (1918), in tracing out the nervous systems of vertebrates, thought that the general plan for all visual systems would be found in the insect eye, but after studying the insect eye, he wrote that the complexity of the nerve structure for vision is even in the insect something incredibly stupendous.

That the visual and tactile senses were related in the evolutionary development of the eye was recognized by Gregory (1967). He suggested that the first retinal images received by the original nervous system were for detecting patterns by touch. Then the compound eye could be a multiplication element, sensing down a single channel, to explore space in a manner similar to a touch probe. A possible application of this relationship is in the tactile-vision system developed by Bach-Y-Rita and Collins (1972). In this system, the image from a video camera was changed to a tactile format by an electro-mechanical array. This tactile image sensed by the skin was then transmitted to the brain, allowing blind individuals using this device to learn to "see."

11 *The Vertebrate Eye*

We say that the function of the eye is vision, but since all photoreception is not vision and not all photoreceptors are eyes, we must consider these broader and narrower terms before delving into our subject proper— the structure and variations of vertebrate eyes and their relation to the ways of life of their possessors.
Gordon Lynn Walls, *The Vertebrate Eye,* 1942.

Vision, of all our special senses, is most important to us, for about 40% of all the sensory information about our world comes to us through the eye. The vertebrate eye and visual system are highly developed, so it is of considerable interest to inquire how the vertebrate eye is structured for imaging and how it functions upon light excitation. The eye has been likened to a camera optical system—the cornea and lens to focus the light and a photosensitive film, the retina, to record the image. But the eye in vision does not function like a camera, for the processes that give rise to the images of our world are quite complex.

The human eye, as in all vertebrates, evolved a refracting-type eye in which an image is formed from the refraction of light by one or more spherical surfaces that separate media of different refractive indexes. The refracting eye has the great advantage that image formation occurs through an integrative action, so that all rays falling on the eye from a given point source are brought to a single point of focus on the retina. Early in the seventeenth century the optics of this type of image formation were described by Johannes Kepler, and how the eye accomplishes imaging was described by Rene Descartes later in the century. In the beginning of the eighteenth century, William Molyneux (1709) of Dublin published the first treatise on optics showing diagrams of the projection of a real image in the human eye. The image is inverted, and its size is inversely proportional to the distance of the object.

The gross anatomy of the human eye is shown in cross section (Fig. 11.1). The eyeball, which is approximately spherical, houses the complete optical and photosensory apparatus. The optical system consists of the cornea, iris and lens. The two refracting structures are the cornea (refractive index 1.336) and the crystalline lens (refractive index 1.437). The primary function of the cornea is to bend the incoming light to form the image on the retina. The lens is used to adjust the focusing of the cornea to accommodate both near and far vision. The lens also acts as a filter by sharply cutting off the far edge of the ultraviolet region at about 360 nm. A variable aperture is provided by a contractile membranous partition, the iris, that regulates the size of the light opening. The pupil is a hole formed by the iris through which light passes to the lens. In dim light, in order to admit more light through the lens, the pupil opens to an extent governed by the activity of the retina. The optical system provides the necessary optics for focusing the light on the retina. The photosensory retinal rods and cones convert the light to electrical signals via the optic nerve, which transmits the image to the brain. The combined optical and sensory apparatus is paired in two symmetrically

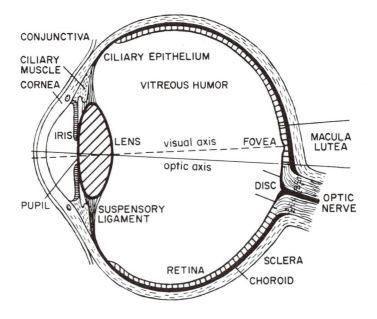

Figure 11.1   The structure of the human eye.

constructed and oriented eyeballs. As a result, a large section of visual space can be imaged on both retinas binocularly, affording stereoscopic vision.

This general description of the major structures of the human eye is common to all vertebrate eyes; all are modifications of a common plan. Excellent descriptions of the anatomy of the eye, its histology and embryology are given by Walls (1942) and Polyak (1957), and its evolutionary development is discussed by Duke-Elder (1958) and Crescitelli (1972).

In the development of the vertebrate eye epithelial cells form only the cornea and the lens, whereas epithelial cells are the basis of the whole eye in invertebrates. The development of the vertebrate eye consists principally in the conversion of the cells in the wall of the optic cup into the retina. While the cells are multiplying, some of them differentiate into the light-sensitive retinal rods and cones, and others into the nerve cells to which the rods and cones are connected. Although the original connection of the optic vesicle with the embryonic brain persists throughout this process (as the optic stalk), nervous connection of the retina with the brain is formed by the outgrowth of nerve fibers from the nerve cells of the retina through the optic stalk into the brain. So the retinal photoreceptor cells possess many features in common with neural cells.

THE RETINA

The human retina is a complex structure of ten cell layers that are closely attached to the pigment epithelium (Fig. 11.2). The retinal photoreceptors, the rods and cones, are arranged in a single-layered mosaic, a rhabdomlike array (Fig. 11.3b), connected with a highly developed system of interconnecting neurons. In the human retina, there are about 1×10^8 retinal rods and 7×10^6 retinal cones. Toward the center of the human retina there is a depression, the fovea, which is the fixation point of the eye and where vision is most acute. It contains mostly cones, of which there are 4×10^3. The rods become more numerous as the distance from the fovea increases. The fovea and the region just around it, the macula lutea, are colored yellow; they contain a plant carotenoid pigment, xanthophyll (Fig. 7.8). Other cell layers of the retina include the external limiting membrane, the external nuclear layer, the external molecular (plexiform) layer, the ganglion cell layer, the nerve fiber layer, and the internal limiting membrane. The nervous cell layers include the rod and cone cells, the bipolar cells, and the ganglion cells.

The first four layers constitute the neuroepithelial layer and are the neurons of the first order. The remaining layers are considered the cerebral portion, where there exists a complex arrangement of nervous elements resembling in structure and function those of the central nervous system; in

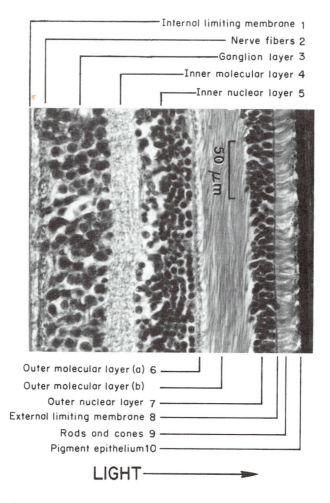

Internal limiting membrane 1
Nerve fibers 2
Ganglion layer 3
Inner molecular layer 4
Inner nuclear layer 5

50 µm

Outer molecular layer (a) 6
Outer molecular layer (b)
Outer nuclear layer 7
External limiting membrane 8
Rods and cones 9
Pigment epithelium 10

LIGHT ⟶

Figure 11.2   The various cell layers in the human retina.

essence the retina represents an outlying portion of the central nervous system. The fifth layer (internal nuclear layer containing the bipolar, horizontal, and amacrine cells) and the sixth layer comprise the neurons of the second order. The seventh and eighth layers make up the neurons of the third order, which pass centripetally to the primary optic center (the lateral geniculate) of the metathalmus. Next to the retina is the choroid coat, a sheet of cells filled with black pigment, which absorbs extra light and prevents internally reflected light from blurring the image.

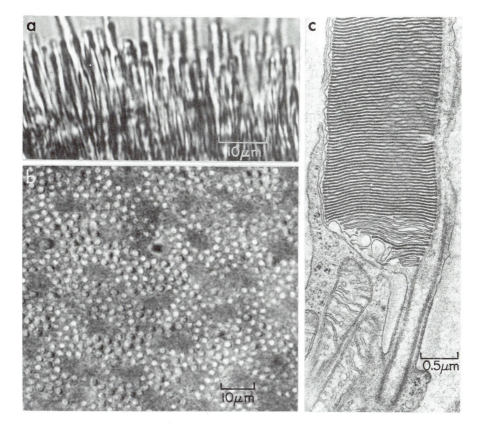

Figure 11.3 The human retinal photoreceptors. *(a)* Retinal rods and cones extending from a bend in the retina. *(b)* Surface view of retinal rods and cones. Note array of rods (rhabdomlike arrangement). From Wolken, 1966, p. 23. *(c)* Electron micrograph of a human retinal rod, showing the outer segment lamellae and inner segment of the retinal cell. (Courtesy of Dr. T. Kuwabara, National Institute of Health, Bethesda, Maryland.)

Light reaching the retina from the lens passes through the nervous cell layers before it strikes the retinal photoreceptors. This inversion of the retina in vertebrates is the result of the development of the eye as an outgrowth of the embryonic brain, whereas in arthropods and celphalopods the eye develops directly from the embryonic skin and the light reaches the photoreceptors directly.

The first description of the structure of the retina is attributed to Leeuwenhoek (1674), but almost two centuries passed before the retinal

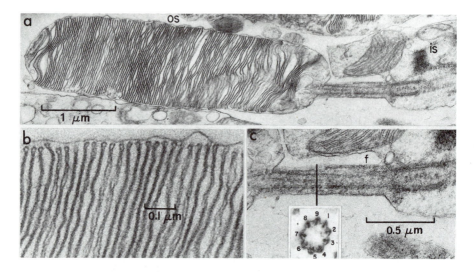

Figure 11.4 Cattle retinal rod. *(a)* Outer segments (os) and connecting flagellum to inner segment (is). *(b)* and *(c)* Enlargements of areas in *(a)*. Electron micrographs.

rods and cones were described by Schultze (1866), and then another decade passed before Böll (1876) recognized that the retinal rods and cones were photosensitive and bleached in the light. Soon afterwards Kühne (1878) and co-workers began a series of studies to isolate the visual pigment. From these observations came the beginning of our understanding of the visual process.

The retinal photoreceptor cells, the rods and cones, are specialized for photoreception. Each cone is connected to a single optic nerve fiber, whereas rods are connected in clusters to optic nerve fibers. Each retinal cell has a rod- or cone-shaped outer segment (OS) that contains all of the photosensitive visual pigment. The inner segment (IS) is the cell body, which contains a nucleus, numerous mitochondria, and other cellular organelles typical of animal cells. The cell membrane of the cell body is continuous with the outer segment. Optically, the index of refraction of the rod (OS) is 1.41 and of the cone, 1.38.

The search for a molecular understanding for visual excitation has led to the structural study of retinal photoreceptors. Developments in electron microscopy and X-ray diffraction have made possible the visualization of the microstructure of the retinal rods and cones all the way down to the molecular level.

Electron microscopic studies have clearly established that the vertebrate retinal rod segments are double-membraned discs, lamellae, of the order of

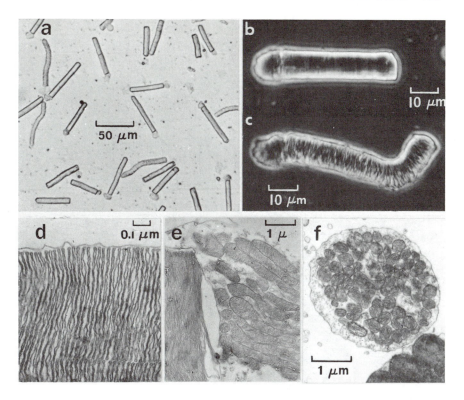

Figure 11.5 Frog retinal rod. *(a)* Freshly isolated rods from retina. *(b)* Phase contrast of a single rod. *(c)* Same rod, bleached at 525 nm. Electron micrographs: *(d)* outer segment; *(e)* and *(f)* inner segment.

250 Å in thickness separated by less dense aqueous proteins and dissolved salts. Each lamellar membrane of the disc is from near 50 to 75 Å in thickness and is composed of a lipid-protein matrix. The visual pigment is complexed to the protein in the membrane. Electron micrographs of the structure of the human, bovine, and frog retinal rods are shown in Figures 11.3-11.6 (refer to schematics of the retinal rod in Figures 11.8 and 12.14). There is no difference in structure among the retinal rods of apes, monkey, cattle, rats, birds, fish and other vertebrates. The retinal cones are also lamellar membranes that are continuous with the cell membrane (Fig. 11.7).

It is curious to note that communication between the outer and inner segments of the rod is accomplished by a cilium or flagellum (Figs. 11.3 and 11.4). The fine structure of this cilium distinctly shows the characteristic nine fibrils found in the cilia and flagella of plant and animal cells (Figs. 6.9 and 6.10). The connecting cilium between these highly specialized outer and inner segments may be a crucial factor in their functional chemistry, for at

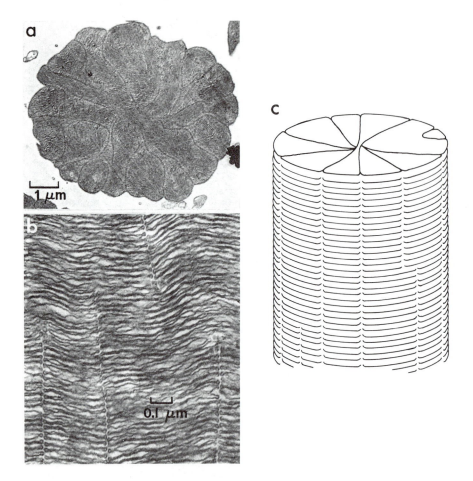

Figure 11.6   Frog retinal rod. *(a)* Cross section of outer segment. *(b)* Longitudinal section. *(c)* Schematic drawing of outer segment of retinal rod. Note structural relationship to fused rhabdom of arthropods. *(After Wolken, 1971, p. 97.)*

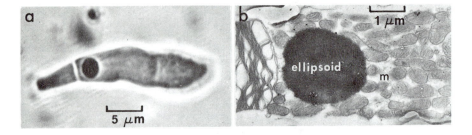

Figure 11.7   Frog retinal cone *(a)*. *(b)* Enlargement of inner segment showing ellipsoid body in the retinal rod cell.

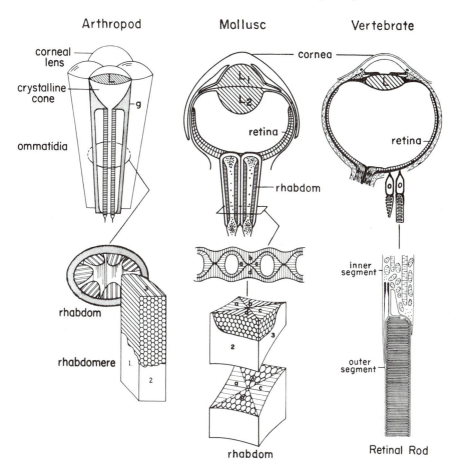

Figure 11.8 Structural development of the eye, from arthropod compound eyes, to molluscs, to the vertebrate eye, showing the optical and photoreceptor structures. *(After Wolken, 1971, p. 142.)*

one end (that of the outer segment) is a highly ordered photosensitive matrix containing the visual pigment, and at the opposite end (that of the inner segment or the cell body) is a mass of tightly packed mitochondria whose enzymatic action provides the chemistry for oxidation-reduction reactions and energy transfer.

EVOLUTIONARY ADAPTATIONS

The structural relationships between the arthropod compound eye, the mollusc eye, and the vertebrate eye are illustrated in Figure 11.8. The phylogenetic evolutionary relationships between animal phyla are shown in Figure 11.9. There are still many unknowns in this phylogeny.

The lowest of the vertebrates are the cyclostomes, which include the dogfish and lampreys. Just above the cyclostomes are the many types of true fish whose eyes are more specialized than those of the cyclostomes. The oldest of the true fish are the elasmobranchs, whose modern representatives

Possible Phylogenetic Relationships between Animal Phyla

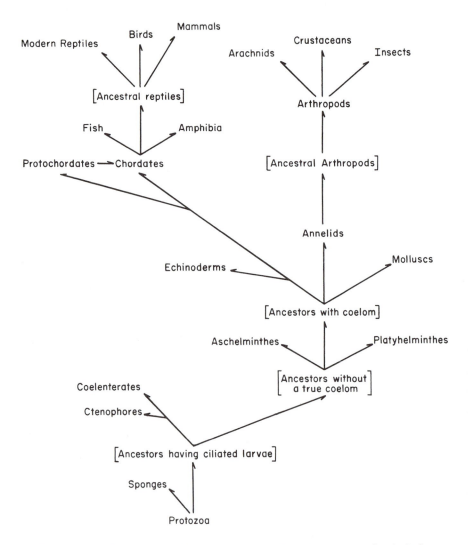

Figure 11.9 Possible phylogenetic relationships between animal phyla.

are the sharks and rays with eye structures resembling the refracting eye of land vertebrates. From one of these creatures evolved the first land animals, the ancient amphibians that gave rise to more developed amphibians and reptiles. The reptiles differentiated into a large number of orders, only four of which have persisted to the present. From one group of extinct reptiles came the birds, whose eyes have many features in common with the eyes of reptiles. Most birds are highly active diurnal animals, and their retinas have numerous single and double cones but relatively few rods and many pigmented oil globules, ranging from colorless to yellow, orange, and red. In nocturnal birds there are a greater number of rods, and the oil globules are colorless. These oil globules are discussed more in relation to color vision in chapter 12.

Peculiar to the eye of birds is the pecten structure, which projects freely into the vitreous of the eye. It is a convoluted and folded structure that varies in size and complexity. The function of the pecten is to improve the visual acuity and detection of moving objects.

Let us look more closely at the structure of the eyes of vertebrates, and in particular to those of amphibians, fish, and birds, whose eyes show evolutionary adaptations to their environment.

The fish eye is similar in structure to the refracting eye of all vertebrates, but the refractive power of the cornea is eliminated under water and the lens is the important structure. The fish lens is spherical, very dense and rigid, with a high index of refraction. In deep sea teleosts, the spherical lens is very large to maximize the light-gathering power and many have duplex retinas (Munk, 1966, 1984).

James Clerk Maxwell (1861) had a lifelong interest in the eye and its optics. He is associated with the tricolor theory of color vision and the optics of polarized light. There is an interesting legend that, while eating kippers, Maxwell contemplated the structure of the crystalline lens of the fish eye. In search of perfect imaging devices Maxwell's investigation of geometric optics led to the beautiful discovery, published in 1853, of the imaging properties of the "fish eye" lens.

Electron microscopic studies of certain deep sea fish lenses show that they vary in density, where the highest density is in the center and decreases toward the periphery, indicating that the lens has a graded index of refraction. A graded index of refraction in the lens would function as a light concentrator in the optical axis, somewhat like a fiber optic. Though these fish have refracting-type eyes, unlike refracting optics the image is formed by reflection.

In the Galapagos Islands there is the "four-eyed" blenny, *Dialommus fuscus,* which frequents the rocks between tides (Fig. 11.10). This fish is amphibious, and its eyes can adjust for both aerial and aquatic vision because it has two distinct optical systems. The cornea is partitioned and the lens so shaped that it refracts light onto the lower part of the retina when in air, and onto the upper part of the retina when in water. As a result, both aerial and aquatic objects are focused simultaneously on different parts of the retina. The combination of the fish's prismatic cornea, and an index of refraction of 1.0 in air, leads to a double image of the world when out of the water (Fig. 11.11). This double view, or binocular of 20-30° overlap, may contribute to depth perception or simply be an evolutionary adaptive development (Stevens and Parsons, 1980).

Another amphibious fish is the mudskipper, *Periophthalmus,* a member of the family of gobies that flourishes in mud flats along the tropical shoreline from Africa, through Southeast Asia, to Japan. Their eyes are retractable stalks for aerial vision in bright sunlight. Their retina is partitioned; in the lower half are primarily cones, while the upper half has only rods for vision in the mud flats.

The elasmobranchs—sharks, dogfish, skates, and rays—cartilaginous fish which are believed to be descendants of Devonian forms that evolved from placodermlike ancestors. Therefore we can look for evidence that the elasmobranch eye and the ultrastructure of the retina developed from the structures of these more primitive creatures and foreshadowed the development of more advanced vertebrate visual structures.

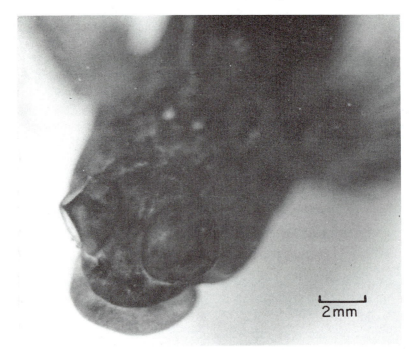

Figure 11.10   Four-eyed blenny fish, *Dialommus fuscus* (Galapagos Islands).

Early anatomical studies of the elasmobranch eye and retina were conducted by Franz (1931, 1934), but surprisingly, there is very little recent research on shark vision. Numerous behavioristic studies of sharks give the impression that the shark is very dependent on vision, although its sense of olfaction is keenly developed (Gilbert, 1963). Most of the available information indicates that the visual cells in the elasmobranchs are undifferentiated and are neither rods nor cones. But the observations by Gruber, Hamasaki, and Bridges (1963) of the lemon shark indicated that the retina has rods and cones. The cones were characterized as being short with tapering outer segments and pyramidal inner segments containing ellipsoids. However, the retinas of some elasmobranchs (e.g., the skate) do not have cones (Dowling and Ripps, 1970).

Most of the visual cells of the shark retina resemble the retina rods of vertebrates. But there are other visual cells that have a greater diameter, and although they have been identified as cones, they appear to be modified rods. Thus, it appears more likely that the elasmobranchs, derived from the placoderms, represent a separate offshoot in the phylogeny of fish. In view of the fact that the sense of olfaction is so well developed in the sharks, one may postulate that their vision has evolved in a particular way to be especially useful in near vision. Once the prey is located by olfaction from a distance, the shark must depend on a fairly acute visual image, which it surely has, even with only one type of visual cell. This visual cell resembles a rod morphologically but, due to the organization of its inner layers, is capable of functioning like a cone. The retina of the smooth dogfish has attained a very

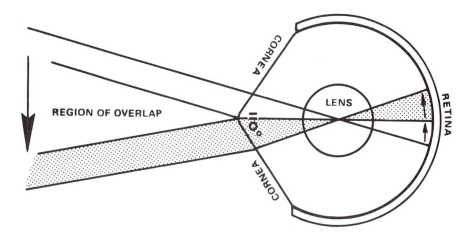

Figure 11.11 Schematic of the optics of the *Dialommus* eye. *(After Stevens and Parsons, 1980.)*

high level of organization comparable to that of the highly visual teleosts and the mammalian eye.

In the amphibian frog eye, the retina and retinal photoreceptors have been extensively studied. The frog retinal rod is an uncommonly large structure of about 60 μm in length and 6 μm in diameter (Fig. 11.5). The retinal rod outer segments can be severed from the retina simply by shaking. Thus they are easily observed with the light microscope. Electron microscopy of the frog outer segments shows that, like the vertebrates, they consist of discs of double membrane lamellae. A cross-sectional view of freshly fixed and sectioned frog rods shows a cylinder with scalloped edges and fissures extending into the rod, so that it is divided into fifteen to twenty irregular wedges (Fig. 11.6). Longitudinal sections reveal that these wedges are rodlets of about 1 μm in diameter within the structure of the rod. In *Necturis,* the mud puppy, the rod's outer segments are about twice the diameter of the frog rod and only one-half its length (12 μm in diameter and 30 μm long). They are the largest known among the amphibians. Here, too, the outer segments are cut by a series of deep longitudinal grooves into an array of rodlets. There are from twenty to more than thirty fissures, making the rod in cross-section appear scalloped as schematically shown in Figure 11.6c.

The structure of the amphibian outer rod segment resembles roughly that of a closed-type rhabdom, as found in the eyes of certain arthropods and molluscs (Figs. 10.7 and 11.8). Perhaps what we are observing in the amphibian rod structure is an evolutionary link with the arthropod and mollusc photoreceptors (Wolken, 1971, 1975). On the other hand, the resemblance between the closed-type rhabdom and the amphibian retinal rod may be the result of independent development and have no evolutionary significance.

REMARKS

We now understand the structure and optics of the vertebrate eye. The question remains how the retinal photoreceptors, the rods and cones, transduce the light energy to electrical signals and how the information is transferred via the optic nerve to the visual cortex of the brain where the images are recorded. These are very complex processes and need to be understood.

12 *Visual Pigments*

The fundamental problem in visual excitation is to
attempt to understand how the action of light on the
visual pigment leads to a nervous response.
George Wald, "Biological Receptor Mechanisms," in
Symposia 16 of the Society for Experimental Biology,
Academic Press, 1962.

What are the visual pigments? How are they synthesized and how do they function upon light excitation? To begin to answer these questions, we must first identify the visual pigment. We have already discussed the fact that carotenoids are easily synthesized by plants and that animals must obtain them from plants.

A discovery of great importance was that it was not the ingested β-carotene, a C_{40}-carotenoid, but its C_{20} derivative, vitamin A, that is necessary for animal life (Fig. 7.11). When the stores of vitamin A in the liver and in the bloodstream have been exhausted, the first symptom of vitamin A deficiency in humans and mammals is the rise of the visual threshold known as night blindness.

The importance of vitamin A aldehyde, retinal (Fig. 12.1), as an intermediate in vitamin A metabolism was established by Morton (1944) and Morton and Goodwin (1944). Then Glover, Goodwin, and Morton (1948) demonstrated that retinal$_1$ is rapidly converted to vitamin A$_1$ when it is administered orally, subcutaneously, or intraperitoneally. The conversion of retinal$_1$ to vitamin A$_1$ is a reduction that occurs in the gut and subcutaneous tissues. This reduction was a plausible explanation for the displacement of the

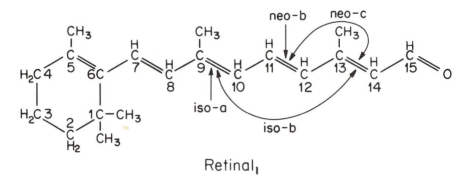

Figure 12.1 Retinal₁ chemical structure, indicating different geometric isomers (see Table 12.1). Refer to Wald, 1953, and Zechmeister, 1962.

Table 12.1   Geometric Isomers of Retinal

Isomerization Around Bonds	Nomenclature	Name
9-10	9-*cis*	iso-*a*
11-12	11-*cis*	neo-*b*
13-14	13-*cis*	nea-*a*
9-10, 13-14	9, 13-di-*cis*	iso-*b*
11-12, 13-14	13, 13-di-*cis*	neo-*c*

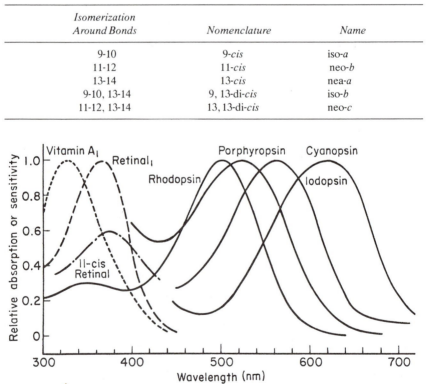

Figure 12.2 Visual pigments, absorption spectra. *(From Wald, 1959, and Dartnall, 1957.)*

absorption maximum (depending on the organic solvent) from around 328 nm, that of vitamin A_1 to around 370 nm, that of retinal$_1$ (Fig. 12.2). The shift in the absorption spectrum is explained by an increase in the number of conjugated bonds from 5 to 6; if the terminal CH_2OH group of vitamin A_1 is replaced by a CHO, the aldehyde group, this change would provide the sixth conjugated bond in retinal.

Retinal has been synthesized from vitamin A_1 by a number of researchers (Morton and Goodwin, 1944; Hawkins and Hunter, 1944). Then it was demonstrated that retinal$_1$ could be obtained by the oxidation of β-carotene (Hunter and Williams, 1945). This discovery is of special interest, for it provided chemical proof of the conversion of β-carotene to vitamin A and demonstrated that retinal$_1$ was an intermediate product in the conversion of β-carotene to vitamin A in vivo. George Wald (1953) established that vitamin A aldehyde, retinal, was the chromophore of the visual pigment, rhodopsin.

The photosensitive visual pigments are then rhodopsins, upon which the visual threshold depends (Fig. 12.3 and 12.4). They are synthesized from aldehydes of vitamin A_1 or A_2, to retinal$_1$ or retinal$_2$ and form a complex with the protein opsin. The extracted visual pigments could be identified by their color and principally by their absorption spectral peaks as rhodopsin (retinal$_1$ + rod opsin) or porphyropsin (retinal$_2$ + rod opsin) for the retinal rods, and

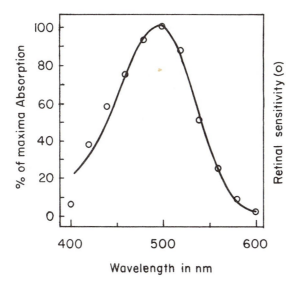

Figure 12.3 Comparison of the absorption spectrum of human rhodopsin to that of the action spectrum for the rod sensitivity of the human eye; circles indicate rod sensitivity. *(After data of Crescitelli and Dartnall, 1953.)* Compare to rhodopsin, Figure 12.2.

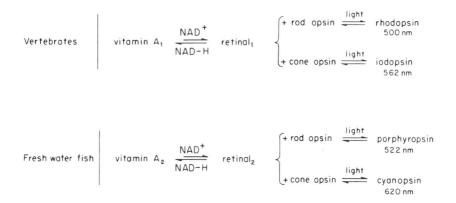

Figure 12.4   Visual pigments in the rods and cones, derived either from Vitamin A$_1$ or Vitamin A$_2$, indicating their absorption maxima.

iodopsin (retinal$_1$ + cone opsin) or cyanopsin (retinal$_2$ + cone opsin) for the retinal cones (Figs. 12.2 and 12.4). Vertebrates and marine fish characteristically possess rhodopsin (retinal$_1$). Freshwater fish possess porphyropsin (retinal$_2$). But fish that migrate between freshwater and marine environments possess both retinal$_1$ and retinal$_2$.

The chemical isomers retinal$_1$ and retinal$_2$ can be distinguished from one another by differences in their spectroscopic absorption peaks.

Vitamin A and retinal are polyene chains that can exist in a number of different geometrical configurations corresponding to the possible *cis-trans* isomerizations around the double bonds of these molecules (Fig. 12.1 and Table 12.1). For example, there are five possible geometric isomers of retinal, corresponding to rotation about the 9-10 carbon bond, the 11-12 bond, the 13-14 bond, the 9-10 and 13-14 bonds, and the 11-12 and 13-14 bonds. The retinal produced by bleaching rhodopsin has always been found to be in the all-*trans* form. However, to resynthesize rhodopsin from all-*trans* retinal, it was found that retinal must be isomerized to the 11-*cis* form in order to recombine with opsin (Fig. 12.5). The 9-*cis* retinal can also complex with opsin to form a series of photosensitive pigments, isorhodopsins, which have been found in small concentrations in the liver and in the blood, but not in the eye.

The existence of the 11-*cis* isomer was considered improbable, because the steric interference between the methyl group at carbon 13 and the hydrogen at position 10 would prevent the molecule from becoming entirely planar. Despite this, Wald and his associates found that the functional

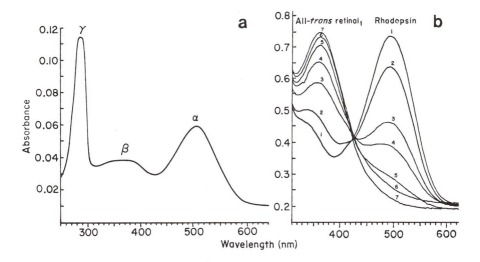

Figure 12.5 *(a)* Spectrum of frog rhodopsin (extracted in 4% tergitol); the absorption peak around 280 nm is due to the protein, opsin. *(b)* Spectra upon light bleaching curves 2–7. Refer to Wolken, 1966*b,* pp. 40–43 and Wolken, 1975, pp. 196–200.

isomer of retinal was in fact the 11-*cis* configuration of retinal. Because of its hindered configuration, the 11-*cis* form is the least stable of the possible isomers; it is the most easily formed upon irradiation, and the most sensitive to light and temperature. What could be more appropriate for vision than a molecule that is very unstable in the light, but stable in the dark?

THE PHOTOCHEMISTRY OF VISUAL EXCITATION

When one goes from dim to bright light, the sensitivity of the eye shifts toward the red end of the spectrum. This observation was first described in 1825 by the Czech physiologist, Johannes Purkinje. The basis of this change is a difference in spectral sensitivity between the retinal rods and cones. In vision the retinal rods function at relatively low light levels, and their maximal sensitivity is in the blue-green at about 500 nm (Figs. 12.2 and 12.3). The retinal cones function in bright light and for color vision, and their maximum spectral sensitivity is shifted more toward the yellow-green at around 562 nm and further in the red of the visible spectrum (Figs. 12.2 and 12.4).

The mechanism of how visual excitation takes place in the retina contin-

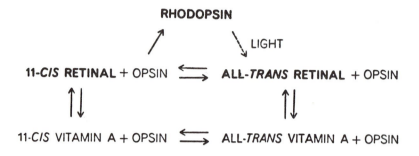

RHODOPSIN

Figure 12.6 Generalized photochemical scheme in light bleaching and regeneration of rhodopsin.

ues to be vigorously investigated. Let us briefly review the photochemistry of the visual pigments. When light strikes rhodopsin in the retinal rods, it bleaches (red → yellow). This effect is accompanied by a shift in the rhodopsin absorption spectrum from around 500 nm to 370 nm. The photochemistry, the bleaching of rhodopsin by light, can be followed by spectroscopy and is illustrated in Figure 12.5, where curve 1 is the spectrum of rhodopsin and curves 2-7 show the displacement of the major rhodopsin peak from around 500 nm to that of all-*trans* retinal near 370 nm. The primary photochemical step then upon light excitation is the isomerization of the 11-*cis* retinal to all-*trans* retinal (Fig. 12.6).

The bleaching of rhodopsin to release retinal from opsin does not proceed directly, but in a series of intermediate steps (Fig. 12.7). The exact number and kinds of intermediates in the photochemistry of rhodopsin are still being explored. The first and only light-catalyzed step is a complicated sequence of events in which the protein-bound retinal forms a high-energy photoproduct that undergoes thermal decay through a series of intermediates. Some of these intermediates have distinct absorption spectra. In the process, 11-*cis* retinal is isomerized to all-*trans* retinal, as it transforms to a more stable form (Menger, 1975), so the only apparent action resulting from the absorption of light by rhodopsin is the isomerization of 11-*cis* retinal to all-*trans* retinal, which releases the energy used in visual excitation. All other reactions can be considered dark reactions. To complete the cycle, the all-*trans* retinal slowly isomerizes back to the 11-*cis* isomer, which recombines with opsin to again form rhodopsin. The reaction is spontaneous, and therefore opsin may be looked upon as a retinal-trapping enzyme, removing free retinal from the mixture, and causing the production of additional retinal from vitamin A to maintain the necessary equilibrium.

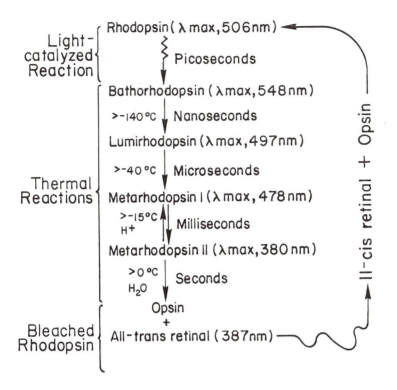

Figure 12.7 Rhodopsin (bovine); photochemical intermediates in the process of light and thermal bleaching reactions of rhodopsin.

OPSINS

Visual spectral sensitivity for different animal species and the rhodopsin extracted from the eyes of these animals show shifts in their absorption maxima, but retinal itself does not influence the absorption maximum of spectral sensitivity. We infer that the visual protein opsin is species-specific, and opsins from different sources complexed with retinal determine the absorption maxima for the different species. There is evidence for this inference in the correlation observed between the habitat of the various species of animals and the spectral absorption maxima of their visual pigments.

A dramatic example of adaptation through modification of protein structure is the relationship between the absorption spectra of the visual pigments of marine fish and the depth of their habitat. Clarke and Denton (1962) observed that the increasing blueness of light with depth in the ocean brings about a shift in the absorption characteristics of the eyes of deep sea fish toward the blue end of the spectrum. This observation was confirmed by comparing the absorption spectrum of rhodopsin extracted from surface fish to that of fish living at depths of 200-400 m. The spectrum is shifted from around 500 nm to around 480 nm towards the blue. Because opsins themselves do not absorb light in the spectral range 480-560 nm, these shifts in the visual pigment absorption maxima are attributed to the effect of the opsins on the spectral properties of rhodopsin.

The molecular weights of opsin from rhodopsins from various animal species also differ. At present we know little about the chemistry and structure of the visual protein opsin from different animal species.

COLOR VISION IN VERTEBRATES

How do animals see colors? Only one kind of visual pigment is sufficient to see, but with only one visual pigment the world is monochromatic; objects appear in black, white, or shades of gray. For animals to see colors requires a minimum of two visual pigments with different absorption sensitivities, and for a full range of color vision at least three (a *trichromatic* visual system) and even four (a *tetrachromatic* visual system) are necessary. The ability to see colors adds a new dimension to experiencing the world.

Cone pigments are for bright light and for color vision. How do they function? It was noted earlier that chemical isolation has yielded only one photosensitive cone pigment, either iodopsin (λ max 560 nm) if retinal$_1$ is present, or cyanopsin (λ max 620 nm) if retinal$_2$ is present.

How do these two pigments fit into an account of color vision in vertebrates? There are two general theories of color vision; one is the tricolor theory, which arose from the work of Young (1802, 1807), von Helmholtz (1852, 1867), and Maxwell (1861, 1890), and the other is the theory of Hering (1885). The tricolor theory asserts that there are three differing pigments in the retinal cones with absorption in the blue, in the green, and in the red regions of the spectrum. According to this theory, the brain computes yellow and white from green and red at high light intensities and white from blue at low intensities. In contrast, Hering's theory postulates that there are the following six basic responses, occurring in pairs: blue-yellow, red-green, and black-white. Excitation leading to any single response suppresses the action of the other member of the pair.

To test these theories, the cones in situ have been isolated and their absorption spectra obtained by microspectrophotometry. For example, frog cones were found to have general absorption throughout the whole of the visible spectrum with maxima near 430, 480, 540, 610, and 680 nm (Wolken, 1966*b*). For the carp, cone absorption peaks in the regions of 420-430,

490-500, 520-540, 560-580, 620-640, and 670-680 nm were found (Hanaoka and Fujimoto, 1957). In the goldfish, which belongs to the carp family, Marks (1963) found cones with absorption peaks at 455, 530, and 624 nm, and Liebman and Entine (1964) found absorption peaks at 460, 540, and 640 nm.

A strong correlation can be found between the absorption peaks exhibited by monkey fovial cones and those found in humans. In monkeys, peaks were found at 445, 535, and 570 nm, closely approximating the human peaks of 450, 525, and 555 nm, and comparing even more closely with the psychophysical data for the spectral sensitivity of the human eye with peaks at 430, 540, and 575 nm (Marks, Dobele, and MacNichol, 1964; MacNichol, 1964; Wald, 1964b; Brown and Wald, 1964; and Wald and Brown, 1965). These studies indicate that for color vision there are at least three spectrally absorbing cone pigments consistent with the psychophysical data, one for sensing blue, one for sensing green, and one for sensing red (Fig. 12.8).

PIGMENTED OIL GLOBULES

In the retinas of amphibians, birds, lizards, snakes, and turtles there are pigmented oil globules (Fig. 12.9). These globules are believed to have evolved a long time ago because they are found in the retina of the sturgeon, belonging to the ancestral fish *Chondrosteans*. However, most fish (with the

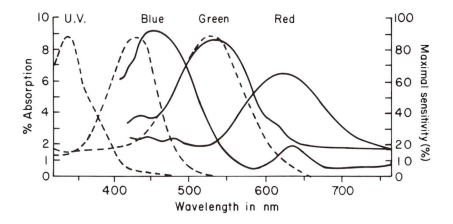

Figure 12.8 Spectral sensitivity of the worker bee eye *Apis mellifera* (- - - -) *(After Autrum and von Zwehl, 1964)*, compared to three different retinal cones of the goldfish eye (——) *(After Leibman and Entine, 1964)*.

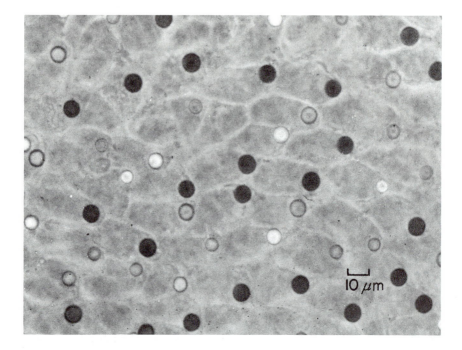

Figure 12.9   Pigment oil globules in the retina of the swamp turtle *(Pseudemys scripta elegans)*.

exception of the lungfish) do not have them. The globules are situated between the inner and outer segments of the cone, so that light must pass through them before reaching the photosensitive visual pigment in the outer segment. The globules are thus in a position to act as color filters.

Over a century ago Krause (1863) suggested that these colored oil globules, by differentially transmitting light to the outer segments, affect the spectral response of the cone, and thus can provide the basis of color differentiation in the animals possessing them. Birds function on relatively high light levels, and cones predominate in their retinas. In freshly excised chicken retinas the pigmented oil globules are "colorless," yellow-green, orange, and red, and range from 3 μm to 5 μm in diameter. They are beautiful to observe in freshly excised retinas. Wald (1948) isolated from the chicken retina three different colored carotenoid fractions that resembled the oil globule colors and identified them as lutein, zeaxanthin, and gallaxanthin.

The absorption spectra obtained by microspectrophotometry of the in situ chicken yellow-green, orange, and red globules show absorption maxima in three different regions of the spectrum (Strother and Wolken, 1960). The yellow-green globules have a general absorption in the region 390-440

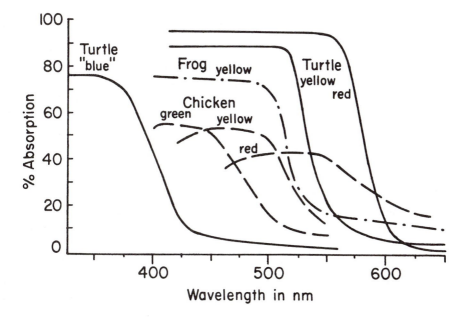

Figure 12.10 Spectra of various colored oil globules acting as retinal filters in the eye of the turtle (———), the chicken (- - - -), and the frog (- · - ·).

nm. The orange globules have a broad absorption in the region 440–480 nm, and the red globules, 480–560 nm (Fig. 12.10). Observations on light bleaching indicated that the yellow-green globules are the most unstable of the three. When stored in the dark at 4°C, they become colorless. When all the colored globules are mixed together, a reddish orange color results, which, when exposed to white light, bleaches first to orange, then to yellow, and then becomes colorless. If one assumes that the globules are acting merely as color filters, then the yellow-green and orange globules are relatively inefficient. For example, the chicken cone pigment, iodopsin, has its absorption peak around 550 nm, and only the red globule has appreciable absorption in this region.

These colored retinal oil globules have been postulated as the basis for the observed increase in spectral sensitivity of these animals to red light as compared to animals without them (Walls, 1942; Fox, 1953). The colorless globules have been identified with the chicken retinal pigment gallaxanthin and could function to screen out damaging ultraviolet light. In the pigeon, the colored oil globules have absorption spectral maxima around 470–490, 540–550, and 600–620 nm, which could account for a shift in the spectral sensitivity toward the red part of the spectrum by the pigeon (Strother, 1963).

For the swamp turtle, the spectral sensitivity electroretinogram (ERG) shows maxima around 575, 620, and 645 nm. Absorption spectra recorded by microspectrophotometry for their colored oil globules show absorption maxima for the red globules from 555 nm to 565 nm, for the yellow globules from 510 nm to 515 nm, and for the colorless globules from 370 nm to 380 nm (Fig. 12.10), confirming the assumption that the spectral response of the turtle eye corresponds to the absorption curve of its cone pigment, cyanopsin. Spectral absorption peaks at 625 nm and 650 nm were also found in agreement with two of the electroretinogram's spectral sensitivity peaks (Strother, 1963; Lipetz, 1984).

In the amphibians, the frog retina has yellow oil globules (Fig. 12.10). The frog also possesses green rods in addition to the red rods, as well as cones. The green-rod absorption peak is near 540 nm and the red-rod absorption peak is near 610 nm, whereas the cones have their maximum absorption near 570 nm. These spectral absorption peaks taken together would cover the spectral range necessary for color vision. The oil droplets then provide a basis for processing color information in the vertebrate retina (Ohtsuka, 1985).

In mammals, the primate fovea, which contains mostly cones, and the region surrounding it, the macula lutea, are colored yellow. The pigment responsible is the plant carotenoid xanthophyll lutein, which could then provide a color filter for the three different spectral absorption peaks of the cones. The yellow oil globule pigment serves as a yellow filter and has basically the same function, that is, to improve the visual acuity, contrast, and hence resolution for all animals that possess such a light filtering system.

In humans the lens yellows with age and therefore the lens could act like the colored oil globules by increasing the acuity of the eye. In this regard, such a light filtering system, together with the retinal cones could provide a tetrachromatic visual system instead of the trichromatic visual system by providing an ultraviolet sensitivity for daylight terrestrial animals.

Some interesting history in the development of color photography is relevant to this discussion of the pigmented oil globules and color vision. In 1894 the French chemist Louis Lumière brought out an autochrome process for color photography (see Lumière and Lumière, 1894). Lumière used suspensions of starch grains dyed blue, green, and red mixed in roughly equal proportions and spread them over the surface of a photographic plate. The granules were squashed flat and the interstices filled with carbon black, so the plate was completely covered. He found that each colored granule served as a color filter for the photosensitive emulsion that lay beneath it, resulting in a colored photograph. Almost simultaneously, Siegfried Garten (1907), an ophthamologist, developed a system of color photography based on the same principle.

INVERTEBRATE VISUAL PIGMENTS

There is good reason to believe that throughout the animal world the kinds of pigment molecules involved in visual excitation are similar. If so, then we must expect certain features of vertebrate visual chemistry to repeat themselves in the invertebrates. Sufficient experimental data has now accumulated to indicate that the invertebrate visual system is indeed similar to the vertebrate visual system; all invertebrate visual pigments are rhodopsins and the chromophore is 11-*cis* retinal$_1$, just as in the vertebrate visual systems (Wolken, 1971; Goldsmith, 1972).

The spectral sensitivity and the concentration of rhodopsin for arthropod and mollusc photoreceptors (Table 12.2) are of the order of 10^7 molecules per rhabdomere, comparable to those of rhodopsin in the retinal rods of vertebrates (Table 12.3). But unlike vertebrate rhodopsin, in the bleaching photochemistry of invertebrate rhodopsins the photo products differ notably from those of vertebrate retinal rods (Fig. 12.7).

COLOR VISION IN INVERTEBRATES

Behavioral studies have shown that the arthropod eye can sense dim light like the rods and bright light like the cones of the vertebrate eye. Also, certain insects and crustacea can distinguish colors. This fact implies that they should have different spectrally absorbing visual pigments, as do the cones in the vertebrate eye. In examining the arthropod visual pigment, two different spectrally absorbing pigments were found that resemble the spectral characteristics of the vertebrate retinal rods and cones (Fig. 12.2). For example, the lobster visual pigment shows absorption peaks at 480 and 515 nm, the crayfish near 510 and 562 nm, the honeybee near 430 and 530 nm, the housefly near 440 and 510 nm, and the moth near 450 and 545 nm. In addition, all exhibit strong absorption in the ultraviolet in the neighborhood of 340-390 nm, which the human eye can not (Loew and Lythgoe, 1985).

To clarify the dual spectral sensitivity and spectrally different absorption peaks, Autrum and Burkhardt (1961) and Burkhardt (1962) measured spectral sensitivity for the blowfly, *Calliphora erythrocephala,* using microelectrodes placed in single retinula cells. Three different spectral sensitivities were found in the visible with maxima at 470, 490, and 520 nm and in the ultraviolet around 350 nm. In similar experiments with drone bees, Autrum and von Zwehl (1962, 1964) found two different receptors with maxima at about 340 and 447 nm. The 447 nm peak compares well with the observed spectral sensitivity maximum for drones at 430 nm, and to the extracted photosensitive honeybee pigment, maximal near 440 nm (Goldsmith, 1958*a*, 1958*b*).

Table 12.2 Some Invertebrate Visual Absorption Maxima

Phyla	*Class*	*Organism*	*Common Name*	*Maxima (nm)*
MOLLUSCA				
	CEPHALOPODA			
		Eledone moschata	octopus	470
		Loligo peallii	squid	493
		Octopus ocellatus	octopus	477
		Octopus vulgaris	octopus	475
		Sepia esculenta	cuttlefish	486
		Sepia officinalis	cuttlefish	492
		Sepiella japonica	cuttlefish	500
		Todarodes pacificus	squid	480
ARTHROPODA				
	CRUSTACEA			
		Callinectes sapidus	blue crab	480
		Euphasia pacifica	shrimp	462
		Homarus americanus	lobster	515
		Leptodora kindtii	water flea	510, 550
		Orconectes virilis	northern crayfish	510, 562
		Palaemonetes vulgaris	prawn	390, 496, 540, 555
		Porcellio scaber	wood louse	480
		Procambaras clarkii	swamp crayfish	570
	INSECTA			
		Apis mellifera	honeybee	440, 510
		Blaberus giganteus	cockroach	495
		Blatta orientalis	cockroach	500
		Calliphora erythrocephala	blowfly	470, 510
		Deilephila elpenor	moth	440-460, 540-550
		Musca domestica	housefly	437, 510
		Manduca sexta	moth	350, 450, 490, 530
		Periplaneta americana	cockroach	500
	ARACHNIDA			
		Evarcha falca	jumping spider	350, 440, 540
	MEROSTOMATA			
		Limulus polyphemus	horseshoe crab (king crab)	520

Source: Based on data from Wolken, 1971, p. 138 and Wolken, 1973, pp. 1269-1270.

In the blowfly eye-color mutant "chalky," which lacks all eye screening pigments, the rhabdom consists of seven rhabdomeres. Langer and Thorell (1966) found that for six of their rhabdomeres the absorption maximum was about 510 nm and for the seventh (the asymmetric) rhabdomere the absorp-

Table 12.3 Concentration of Visual Pigment, Rhodopsin,
in Arthropods and Vertebrates

		Average volume in cm^3	*Concentration of rhodopsin molecules*
Arthropods		rhabdomeres	
Cockroach	*Periplaneta americana*	3.1×10^{-10}	5.9×10^7
Cockroach	*Blatta orientalis*	3.0×10^{-10}	4.3×10^7
Housefly	*Musca domestica*	6.8×10^{-11}	3.7×10^7
Honeybee	*Apis mellifera*	2.6×10^{-11}	2.6×10^7
Vertebrates		retinal rod	
Amphibian			
Frog	*Rana pipiens*	1.5×10^{-9}	3.0×10^9
Cattle		7.5×10^{-12}	1.0×10^6
Human		1.6×10^{-10}	1.0×10^7

Source: Refer to Wolken, 1971, p. 150.

tion maximum was about 470 nm. These spectral peaks come close to Burkhardt's maxima for spectral sensitivity.

Presumably these absorption peaks are associated with two differently absorbing visual pigments. To find that there are two differently absorbing pigments in different rhabdomeres of the same rhabdom raises the interesting possibility that these different rhabdomeres could function as do rods and cones in the vertebrate eye.

SCREENING PIGMENTS

The various eye colors seen in the arthropods come from differently colored pigment granules which surround the ommatidia and regulate the light that reaches the rhabdom. It was thought that the arthropod pigment granules would have similarities to the pigmented oil globules found in the retinas of birds, turtles, lizards, and snakes (Wolken, 1971), but there is no evidence to indicate that they are chemically identical.

In the compound eye of the housefly, *Musca domestica,* yellow pigment granules are found at the top of the ommatidium, close to the corneal lens and crystalline cone, whereas the red pigment granules are found closer to the rhabdom. Microspectrophotometry of the yellow and red granules showed that the yellow pigment in these granules has a broad absorption band maximal around 440 nm; the red pigment has a broad absorption band with its maximum around 530 nm and smaller peak near 390 nm (Strother and Casella, 1972). The yellow pigment corresponds to the plant pigment xanthophyll, found in the fovea of the mammalian retina. The red pigment exhibits a shift in absorption maximum from about 490 nm to 440 nm in changing from alkaline to acid conditions, which is similar to the shift shown by rhodommatin, a red ommochrome-type pigment isolated from insects (Bowness and Wolken, 1959; Butenandt, Scheidt, and Bickert, 1954).

In the blowfly, *Calliphora erythrocephala*, the yellow pigment granules' absorption peak is around 445 nm and the red pigment granules have absorption peaks near 380 and 540 nm (Langer, 1967). These absorption peaks are similar to the housefly whose yellow pigment was identified as an oxidized xanthommatin and red pigment as a rhodommatin.

These spectra are something of an anomaly, for no obvious correlation between them and the overall spectral sensitivity for these insects is evident. Accordingly, Langer and Thorell (1966) made direct microspectrophotometric measurements of the blowfly rhabdomeres within a single rhabdom. They found two different spectra; one had two absorption peaks with maxima at about 380 nm and 510 nm and closely resembled the red screening pigment spectrum, and the other showed only a single peak near 470 nm and its shape closely resembled the yellow pigment spectrum (Strother and Casella, 1972). These results would indicate that the screening pigments function to absorb light for the visual pigments.

However, it is possible that the yellow and red screening pigments are acting separately to screen two different visual pigments, namely, the one absorbing at 440 nm and the other near 510 nm. This possibility suggests that those spectral regions, where the pigments permit passage of light, coincide with the regions where the insects are strongly sensitive. Thus, ultraviolet light of 350-400 nm and red light beyond 600 nm are transmitted and not absorbed by the screening pigment granules. This transmitted light is then available to the photoreceptors and results in greater visual sensitivity for these insects.

THE UNIVERSALITY OF RETINAL

The visual pigment rhodopsin, and hence its chromophore retinal, is found in all animals with eyes. Until recently retinal was not found in eyeless animals, and it was believed that only animals could synthesize retinal from its precursors, β-carotene and vitamin A. Therefore it was of interest to search for evolutionary clues to the biosynthesis of retinal and the visual pigment rhodopsin in primitive photoreceptor systems like those of the bacteria, fungi, algae, and protozoa.

Phototropism and phototaxis were discussed in chapter 9. In examining phototropic behavior of the fungus *Phycomyces,* the question was raised, "What is the photoreceptor molecule?" Experimental evidence indicated that a flavoprotein is the photoreceptor molecule and that carotenoids were also linked to the photoprocess. Among the numerous carotenoids extracted from the sporangiophores, retinal was identified (Meissner and Delbruck, 1968), suggesting that a rhodopsinlike molecule could be the photoreceptor pigment. From our studies, the most likely place to search for retinal was in the octahedral crystals found in the light-growth zone (Fig. 9.5). A test to identify retinal is the reaction with the Carr-Price reagent ($SbCl_3$ in chloroform), which gives a maximum absorption peak at 664 nm (Hubbard, Brown, and Bownds 1971). The octahedral crystals were isolated from stages (I-III, and when reacted with the Carr-Price reagent, gave an absorption peak at 663 nm, confirming the presence of retinal. Another related test was to take the crystals, fix them in 4% gluteraldehyde, and measure their absorption spectrum. The result this time showed a maximum near 520 nm, indicating the presence of a rhodopsin. Whether a retinal-rhodopsin type system is the photoreceptor for *Phycomyces* phototropism or whether it functions in the light-growth response is a question awaiting further studies.

This evidence brings us to the discovery in the past decade of retinal in the extremely halophilic bacteria. These bacteria grow in environments with an unusually high salt (NaCl) concentration of 25%, at high temperatures near 44°C, and in continuous direct sunlight. They synthesize large quantities of carotenoids that give the culture a red to purple color. A surprising finding for the bacterium *Halobacterium halobium* was that the ruptured membrane described as the "purple membrane" possesses the visual pigment rhodopsin (Oesterhelt and Stoeckenius, 1971; Blaurock and Stoeckenius, 1971). The purple membranes can be easily isolated and their absorption spectrum resembles the vertebrate cone pigment iodopsin (Fig. 12.11*a*), for it has its maximum absorption in the visible at 560 nm and in the ultraviolet at 280 nm. The pigment was named *bacteriohodopsin* and has properties similar to vertebrate rhodopsin. The chromophore was identified at retinal$_1$, which is complexed with an opsin through a Schiff-base linkage with the amino group of lysine, in a manner similar to vertebrate rhodopsin.

To determine whether bacterial retinal would form rhodopsin when combined with the visual protein opsin of animal eyes, opsin was prepared from cattle retinas because bovine opsin is one of the more stable of the opsins (Hubbard, Brown, and Bownds, 1971). The cattle opsin absorption spectrum showed that it resembles other opsins, with its major absorption peak around 280 nm (Fig. 12.11*a, b*).

Because the 11-*cis* isomer of retinal will complex with opsin to form rhodopsin, we took the bacterial retinal in a 1.8% digitonin solution in buffered phosphate, pH 6.3 M/15, and incubated it with an excess of cattle

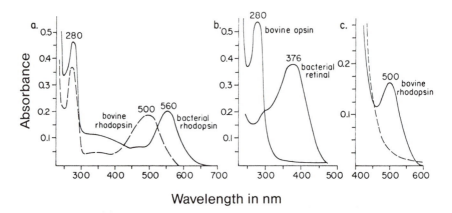

Figure 12.11 *(a)* Absorption spectra of bacterial rhodopsin *Halobacterium halobium* and bovine rhodopsin. *(b)* Absorption spectra of bacterial retinal and bovine opsin. *(c)* Bovine rhodopsin formed upon incubation of bacterial retinal and bovine opsin, and after irradiation (- - - -).

opsin for 3 hours in the dark at room temperature. This experiment resulted in a shift of the absorption spectrum maximum peak to 500 nm, that of cattle rhodopsin (Fig. 12.11*c*). The rhodopsin formed was photosensitive, and the peak at 500 nm bleached when irradiated with light. The fact that bacterial retinal will complex with vertebrate opsin to form cattle rhodopsin demonstrates that the protein opsin is species-specific (Wolken and Nakagawa, 1973). This specificity holds true for all animal visual pigments.

What function does rhodopsin perform in these bacteria? It was found that the bacterium shares with both plants and animals some features of the system it uses to exploit light energy. Normally, the bacterium synthesizes adenosine triphosphate (ATP) by oxidative phosphorylation. Unlike animal visual systems, the retinal of bacteriorhodopsin is in the all-*trans* configuration and light causes it to isomerize to the 13-*cis* isomer. During a cylical chain of events in which bacteriorhodopsin eventually reverts to its starting form, protons are pumped from the inside to the outside of the cell. The proton gradient thus formed—like the one generated in plant photosynthesis—supplies the energy for ATP synthesis. Under anaerobic conditions in the dark, the ATP of the bacterial cells decreases significantly and light is required to restore the original ATP concentration. But in the absence of light, oxygen is required. Thus halobacteria have adapted to their environment by incorporating rhodopsin in their cell membrane. Rhodopsin transduces the light energy to chemical energy, similar to the function of chlorophyll in the chloroplast membranes, and provides an alternative oxidative mechanism for energy production (Mitchell, 1966).

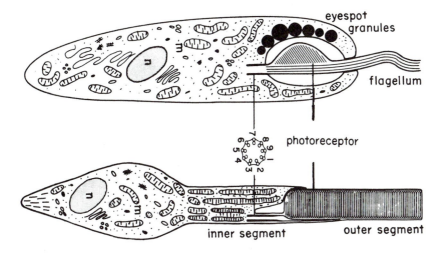

eyespot
granules

flagellum

photoreceptor

inner segment

outer segment

Figure 12.12 A structural comparison between the photoreceptor structure of the *Euglena* cell with that of the vertebrate retinal rod cell.

In *Euglena*, the photoreceptor molecule for phototaxis was identified as a flavoprotein. *Euglena* synthesizes a number of carotenoids in which β-carotene, the precursor molecule of retinal, is found in the eyespot granules. Experimental evidence indicates that β-carotene, or possibly retinal, functions as a light filter for the photoreceptor molecules in phototaxis (Wolken, 1977). However, in the algal flagellate *Chlamydomonas* a retinal-rhodopsin photoreceptor system was identified with the eyespot for phototaxis (Foster and Smyth, 1980; Foster et al., 1984). Similarities can be seen when one compares the photoreceptor structural system with these algal flagellates to a vertebrate retinal rod cell (Fig. 12.12).

The fact that retinal is found in organisms without eyes suggests that it has (as does vitamin A) a more generalized biological function than only vision. This suggestion is the result of evolution of a type of molecule that was synthesized very early in the history of life. The original function of retinal may have been to protect DNA in the cell against radiation damage from the ultraviolet radiation present in the early environment. In the halophilic bacteria, a bacteriorhodopsin functions in the metabolic processes of photophosphorylation similar to that of chlorophyll in photosynthesis, while in the algal flagellate *Chlamydomonas*, with an eyespot structure, it functions for phototaxis. Once retinal became incorporated into the eye for vision it was then used for all visual systems (Fig. 12.13), demonstrating natural selection at the molecular level. Therefore the role of retinal in living organisms is far more universal than was previously thought.

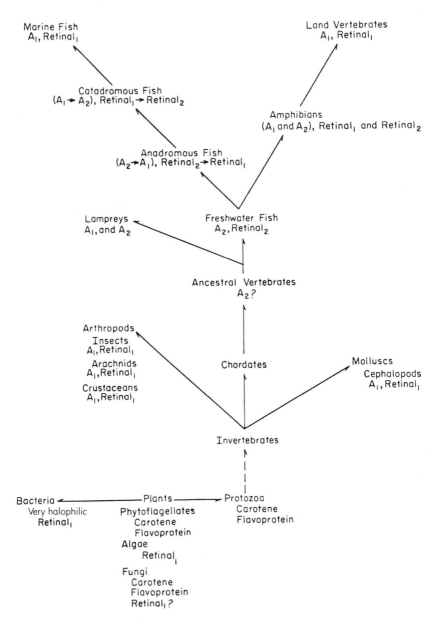

Figure 12.13 Phylogenetic relationship of the visual pigment, Vitamin A, retinal, carotenes, and flavoproteins in various animal photoreceptors and eyes. *(After Wald, 1940 and 1970, and Wolken, 1975, including recent data.)*

RETINAL ROD MOLECULAR STRUCTURE

With the information we have on the visual pigments, their photochemistry, and the structure of the retinal photoreceptors of the eye, let us see if we can visualize a molecular model for the retinal rod outer segment in which the rhodopsin molecules reside. It is apparent from Figures 11.3 through 11.5 that the rod outer segments are composed of double-membraned lamallae with a repeated unit of about 250 Å. Retinal is bound to the amino acid lysine in the protein opsin within the membranes, oriented so it is nearly parallel to the surface of the membrane. When light strikes the retinal rod, the molecule retinal changes its shape from the 11-*cis* to the all-*trans* configuration, and opsin changes shape as well. The conformational shape changes in the opsin set off biochemical processes in the retinal cells that lead to electrical signals.

To develop a molecular model of the retinal rod, the geometry of individual rods and the number of lamellar membranes and their dimensions were determined (Wolken, 1975). With these data and the rhodopsin concentration per retinal rod we can calculate the area that each rhodopsin molecule would occupy on the surface of a membrane. First, though, we must make clear certain assumptions: that the outer segment lamellae are double membranes of lipid and protein, that there is a monomolecular layer of rhodopsin molecules associated with the protein (or lipoprotein) membrane, and that these double membranes are separated by water. All these assumptions are supported by chemical analyses showing that, in general, the visual pigment rhodopsin accounts for 4-10%, protein for around 60%, and total lipids around 40% of the weight of the outer rod segment (Table 12.4).

The cross-sectional area that would be associated with each rhodopsin

Table 12.4 Comparative Composition of Proteins and Lipids in Retinal Outer Segments

	Percent dry weight	
	bovine	*frog*
Total lipid	38.15	40.6
Total protein	61.85	59.4
	Percentage of total lipids	
Phosphatidylethanolamine	38.5	25.2
Phosphatidylserine	9.2	9.5
Phosphatidylcholine	44.5	49.4
Sphingomyelin	1.3	1.8
Other phospholipids	6.5	9.2

molecule is expressed by $A = \pi D^2/4P$, where D is the diameter of the retinal rod and P is the number of rhodopsin molecules in a single monolayer. In our calculation, P is replaced by $N/2n$, where N is the rhodopsin concentration in molecules per retinal rod and n is the number of double-membraned lamallae per rod. So the expression for the maximum cross-sectional area for each rhodopsin molecule is given by $A = \pi D^2 n/2N$. Using this equation, the cross-sectional areas calculated for cattle and frog rhodopsin were 2500 and 2620 Å^2 respectively (Table 12.5), which means the diameter of the rhodopsin molecule should be approximately 50 Å. This value agrees well with theoretical calculations asserting that a rhodopsin molecule, if symmetrical, would have a diameter about 40 Å. Thus, our figure for the available area implies that there would be sufficient space for all of the rhodopsin molecules to cover all of the lamellar surfaces of a single rod for maximum light absorption. The molecular structure for a rod outer segment is schematized in Figure 12.14, and a small area is enlarged to show the molecular packing of rhodopsin in the membrane.

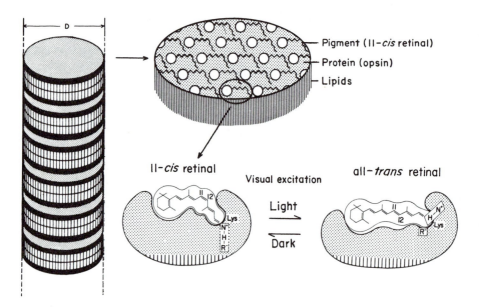

Figure 12.14 Schematized model for the molecular structure of the vertebrate retinal rod, showing the possible molecular geometry of retinal with opsin (rhodopsin) in the membrane of the retinal rod. Note the conformational change upon light absorption that releases the 11-*cis*-retinal from rhodopsin to the all-*trans*-retinal. *(After Wolken, 1975, p. 211.)*

Table 12.5   Retinal Rod Structural Data

Animal	Average diameter, D (μm)	Thickness of dense lamellae, T (Å)	Number of dense layers per rod, n	Number of rhodopsin molecules per rod, N	Calculated cross-sectional area of rhodopsin (Å²)	Calculated diameter of rhodopsin molecule,[c] (Å)	Calculated molecular weight, M[a]
Frog	5.0	150	1000	3.8×10^{9}[b]	2620	51	60,000
Cattle	1.0	200	800	4.2×10^{6}	2500	50	40,000[c,d]

Source: After Wolken, 1961a, p. 51 and Wolken, 1975, p. 210.

[a]Heller, 1969, finds that the molecular weight of cattle and frog rhodopsin is 28,000.

[b]Microspectrophotometry of frog rod gives 3.0×10^{9} rhodopsin molecules.

[c]Calculation based on a lipoprotein, density 1.1, gives a molecular weight of 32,000.

[d]Abrahamson and Fager, 1973, indicated molecular weight of 35,000-37,000.

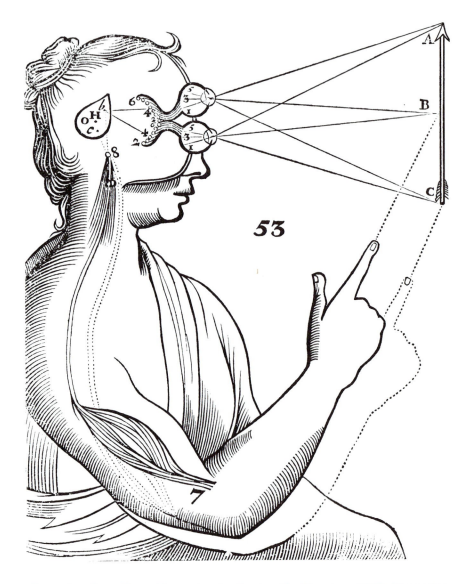

Engraving from Rene Descartes in *Tractus De Homine,* 1677. Descartes assigned the seat of the rational soul to the human pineal (H), in which the eyes perceived the events of the world and transmitted what they saw to the pineal by way of "strings" in the brain. *(Photograph of wood engraving furnished by the Osler Library, McGill University, Montreal, Canada.)*

13

Light That Controls Behavior

Man is, among other things, a remarkably living
sun dial.
Lyall Watson, 1973.

The ability to respond to stimuli is one of the characteristics of living things which appeared early in the course
of evolution, almost as soon as the aggregation of
large molecules to form a cell.
E. Newton Harvey, 1960. *Comparative Biochemistry,*
vol. 2.

EXTRAOCULAR PHOTORECEPTION

How do eyeless and blinded animals sense light? Eyes are not the only means
of photoreception in animals, for their general body surfaces are especially
light sensitive. Behavioral responses to light, or a set of responses that are
initiated through photoreceptors closely bound to dermal and neural cells
(including the brain), are considered extraocular. Research into extraocular
photoreception for all types of photobehavior that was *not* initiated through
the eye was pioneered by Steven (1963) and Millott (1968, 1978). Extraocular
photoreception is of considerable interest for it plays an important role in the
behavior of animals. Let us examine some of these photobehavioral responses
and see how they function, either alone or in conjunction with hormonal
systems and with the visual system.

Extraocular photoreception for the most part is observed as phototactic
movement to or away from the light source. The movement may be in the
bending or contraction of a part or all of the body. Blind cave-dwelling
animals distinguish between light and dark by means of light-sensitive receptors in their skin, which trigger impulses to their muscles, causing them to

seek darker places. This diffuse photosensitivity over the whole or part of an animal's body is recognized as the dermal "light sense."

Many animals respond to a sudden change in light intensity by a withdrawal reaction observable in numerous marine animals and referred to as the shadow response. The shadow response of the sea urchin *Diadema* is a withdrawal from the light and is accompanied by a complex spine waving reaction (Millott, 1968). The marine worm *Golfingia gouldi* reacts to light with a retraction of the proboscis (O'Benar and Matsumoto, 1976). The burrowing sea anemone *Calamactis praelongus* bends toward the light (Marks, 1976). The adult sea squirt *Ciona intestinalis* orients in the direction of the light, accompanied by the opening of its siphons (Dilly and Wolken, 1973). The *Ciona* body surface is also photosensitive to changes in light intensity, and the most sensitive area is found in the region of the ganglion cells. The response to changes in light intensity is localized contraction and elongation of the body. In eyeless bullfrog larva extraocular photoreceptors serve for the perception of celestial cues and for spatial orientation.

The extraocular photoreceptors not only inform the animals of the presence of light but also measure light intensities and select specific wavelengths of light for function.

PHOTOPERIODIC AND RHYTHMIC BEHAVIOR

Light influences rhythmic behavior in living organisms from microorganisms to humans. Many organisms show rhythmic behavior patterns regulated by periods of light and dark, a behavioral phenomenon known as photoperiodism. Although related in some way to photosynthesis in plants and to vision in animals, photoperiodism is independent of them. Pioneering and extensive studies of photoperiodism were made by Bünning (1958, 1973, 1986). Bünning referred to the period in light as *photopic* and in the dark as *scotopic*. These terms were already in use for the visual photoreceptors, referring to the retinal cones that function at high light levels and the rods that function at low light levels, implying that there were equivalent photoreceptors active for photoperiodic behavior.

It had long been supposed that the alternating periods of light and darkness impose the diurnal rhythm. But, if rhythm is related to environmental periodicity, it should disappear when the organism is placed in continuous darkness. Experiments show that the rhythm continues in darkness, that it is controlled from within the organism, and that it is endogenous.

Endogenous clocks are called *circadian* when the natural period is approximately 24 hr (usually between 22 and 26 hr). A change in the pattern of external stimuli results in a shift in phase of the endogenous rhythm. When the plant or animal is raised under constant conditions, a single stimulus, such as a transition from continuous darkness to continuous light, or a short period of light interrupting continuous darkness, is necessary to evoke the periodicity. A signal from the outside, usually the onset of dawn or dusk, keeps the clock in step with the natural day-night cycle. An endogenous rhythm in organisms suggests some clocklike mechanism and is commonly referred to as the biological clock. For example, flies reared in total darkness for many generations still show diurnal rhythms. But if they are subjected to a flash of light in their dark world, they will all resynchronize their rhythms, a phenomenon common to most animals (Dethier, 1962).

The clocks are not necessarily daily clocks. Some are internal timers of less than 24 hr while others are longer timers, perhaps set by lunar periods as in the reproductive cycles of the marine worm *Odontosyllis,* and in humans (a 28-day cycle). Even seasonal clocks exist in some organisms.

Rhythmic migration is a behavioral response associated with extraocular photosensitivity and is evident in many marine animals. This rhythmic behavior coincides with the timing of the tides. Associated with the tidal rhythms are color and pattern changes in the skin of certain animals (Fingerman, 1960). The planarian *Convoluta roscoffensis,* peculiar to the coast of Brittany, France, exhibits tidal rhythm (Keeble, 1910). *Convoluta* buries itself in the sand at night and during high tides, but emerges onto the surface at low tides during the daytime. If the animals are brought into the laboratory and placed in constant light, vertical migration continues for up to 7 days in synchronization with the tides. The rhythm persists only in constant light and not in darkness (Palmer, 1974; Wolken, 1975). In the sea hare, *Aplysia,* daily rhythmic behavior can be modulated by the eye, but is not dependent on the eye. Its extraocular system can directly mediate the response of the circadian oscillator, via the sixth abdominal ganglion (Block and Lickey, 1973; Lickey and Zack, 1973). Flight rhythms in insects are also controlled by extraocular photoreceptors (Dumortier, 1972).

REPRODUCTIVE DEVELOPMENT

The extraocular photoreceptor system also functions in the control of hormonal stimulation of growth and in reproductive development. For example, in the garden slug *Limax maximus,* the extraocular receptor measures the duration of increasing daylight. The increasing day length triggers the secretion of a maturation hormone by the brain, which in turn initiates reproductive development (McCrone and Sokolove, 1979). The influence of light on hormonal-reproductive behavior is also seen in insects; for example, in the pupae of some species termination of diapause will occur when exposed to a photoperiodic regime.

For the vertebrates, amphibians, reptiles, birds, and mammals, extraocular photosensitivity is found to function in reproductive sexual cycles (Wurtman, 1975). The extraocular system can function alone or together with the visual system, but the control of the gonadal responses functions without the visual system (McMillan, Elliot, and Menaker, 1975*a*, 1975*b*). In the human brain the pituitary gland plays a major role in daily and seasonal cycles. The pituitary regulates the hormonal output of other glands and through them exercises control over a wide variety of metabolic and sexual reproductive processes. The pineal gland regulates hormonal reproductive rhythmic behavior.

Extraocular photobehavioral responses are for the most part simple reflexes, but the mechanisms that involve neural, brain, and hormonal systems are very complex.

THE PHOTORECEPTORS

The behavioral responses of extraocular photoreception are now recognized, but the locations and structural identities of the photoreceptors have been more difficult to determine. The extraocular photoreceptors may be different from the retinal photoreceptors of the eye. In many animals the photosensitive areas are widespread and are not equally sensitive or confined to certain areas of the skin.

In the gastropod mollusc *Onchidium,* the photoreceptors are in the periphery of the mantle (Hisano, Tateda, and Kubara, 1972*a*, 1972*b*). In the sea urchin *Diadema,* the skin is pervaded by nerve fibers that are extremely photosensitive (Millott, 1978). In annelids, such as the polychaete *Nereis diversicolor,* the photosensitive area is located on the parapodia and on the pro- and peristromium. In vertebrates the dermal photosensitivity is located in the skin for immature animals, for example, embryos, chicks, and newborn rats (Harth and Heaton, 1973). In the frog, the dermal response to light has been obtained from sections of the frog skin and the skin over the frontal organ (Eldred and Nolte, 1978).

Extraocular photoreception is found in neural tissue cells among a wide range of animals. For example, a receptor site is located in the cerebral lobes of the alfalfa weevil *Hypera postica* (Meyer, 1977). Another receptor site is found in various ganglia for the crayfish *Procambarus clarkii* (sub- and supra-oesophageal and sixth abdominal ganglia) and the scorpion *Heterometrus fulvipis* (Geethabali and Rao, 1973). In the hardshell clam *Mercenaria,* a pigmented structure is located in the distal processes of the axons that is organized into a pentalamellar structure (Weiderhold, MacNichol, and Bell, 1973). Within the photosensitive region of the sea hare *Aplysia,* in the sixth abdominal ganglion neural cells, there are yellow-orange pigmented granules, *lipochondria,* that are membrane-bound, crystalline in structure, and light-sensitive. The lipochondria granules, when illuminated for about 30 sec, undergo a structural change accompanied by the release of calcium (Kraughs, Sordhal, and Brown, 1977; Brown, Baur, and Tully, 1975). The *Aplysia* peripheral nerve, the rhinophore, is also an extraocular photoreceptor site that controls the locomotor and circadian rhythmic activity (Chase, 1979).

In cephalopod molluscs, the squid and the octopus, the photoreceptors are the parolfactory vesicles and the stellate ganglion. The parolfactory vesicles are found near the optic tract and are connected to the brain in parallel with the eye. Their rhabdomlike structure is similar to that of the visual photoreceptor structures of arthropod compound eyes (Baumann et al., 1970; Mauro and Sten-Knudsen, 1972; Wolken, 1971).

The passage of light through the skull to the brain was not considered important in extraocular photoreception in animals until it was demonstrated that light does indeed penetrate through the skull into the brain (Van Brunt et al., 1964). In birds, frogs, lizards, salamanders, and sharks the pineal organ and the parapineal region lie near the surface of the brain and are light-sensitive. The parapineal region (parietal eye in the lizard, stirnorgan in the frog) is connected to the pineal.

Regions of the brain other than the pineal, such as the hypothalamus, pituitary, and rhinencephalon, are also sites for extraocular photoreception (Hisano et al., 1972*c;* Van Veen, Hartwig, and Muller, 1976). In the rat, the photoreceptor lies in the Harderian gland, which is closely connected with deeper brain structures. In blind rats, the circadian rhythm continues, but is interrupted upon removal of the Harderian gland (Wetterberg, Geller, and Yuwiler, 1970; Wetterberg et al., 1970). For mammals, an extraocular photoreceptor that influences circadian rhythms is the pineal.

Obviously, the photosensitive areas and photoreceptors are varied and widely distributed throughout an animal's body (Table 13.1).

SPECTRAL SENSITIVITY AND THE PHOTORECEPTOR

Extraocular photoreception has been primarily determined from behavioral and electrophysiological studies. The wavelengths of light capable of producing the response have been found to be in the blue from about 400 nm to 500 nm, but photo-responses also occur in the near-ultraviolet and infrared regions of the spectrum.

For most insects, the behavioral action spectrum sensitivity peak is around 450 nm (Truman, 1976). Spectral sensitivity in the red region of the spectrum has been noted in the butterfly, with a response near 610 nm (Meyer, 1977) and in the alfalfa weevil, *Hypera postica,* with a response from 650 nm to

Table 13.1   Some Selected Response Data of Extraocular Photosensitivity

Species	Response	Location	Receptor	Spectral Sensitivity λ max
Fruitfly Drosophila melanogaster	eclosion (pupae-adult) phototaxis	brain	nervous tissue	~460 nm (blue region)
Alfalfa weevil Hypera postica	phototaxis	brain	nervous tissue	650 nm 900 nm
Arachnid Choborus americanus	50% diapause termination	brain	nervous tissue	450 nm
Scorpion Heterometrus fulvipes	electro-physiological mediates	telsonic nerves	nervous tissue	568 nm 440 nm
Squid Todarodes sagittatus	migration and slow diurnal rhythms	parolfactory vesicles	nervous tissue	~500 nm
Crayfish Procambarus clarkii	locomotor reflex	6th abdominal ganglion	nervous tissue	502 nm
Pulmonate snail Onchidium verruculatum	shadow reflex regulation of diurnal activity	sub and supra-oesophageal	nervous tissue	white light
Marine worm Golfingia gouldii	contraction of proboscis retractor muscle	brain	nervous tissue	500 nm
Sea squirt Ciona intestinalis	spawning, siphon movement (adult)	gonopore tip neural gland	dermal nervous tissue	415 and 550 nm 460 nm
Tiger salamander Ambystoma tigrinum	orientation to photic cues	brain	pineal	sun light
Lizard Sceloporus olivaceus	rhythm activity entrainment	brain	pineal, parietal eye and other brain region	white light
Southern cricket frog Acris gryllus	orientation to photic cues	brain	pineal and diencephalon	sun light
Dogfish shark Scyliorhinus caniculus L.	inhibition of spontaneous response	brain	pineal	~500 nm
Rock pigeon (chick) Columba livia	head wagging leg extension	skin	possibly dermal	500 W photoflood lamp

Source: After Wolken and Mogus, 1979, p. 190.

farther into the red. In the nematode *Chromadorina viridis,* the spectral sensitivity peak is around 366 nm, in the near-ultraviolet. The pigment responsible for this sensitivity is embedded in the esophageal musculature of *Chromadorina* and is identified as a heme or heme derivative (Croll, 1966). In two species of scorpions, *Heterometrus fulvipis* and *heterometrus gravimanus,* where the pigmented tail segment and telsonic nerves are photosensitive, spectral sensitivity was found to have two peaks, one around 586 nm and the other near 440 nm (Geethabali and Rao, 1973).

In the sea hare, *Aplysia,* the light-sensitive neurons have pigment granules, *lipochondria,* whose major absorption peak is around 490 nm, resembling that of a carotenoid, and a second pigment with an absorption peak around 579, resembling that of a cytochrome (Austin, Yai, and Sato, 1967). Other researchers have found a carotene-protein with absorption peaks at 463 and 490 nm, and a heme-protein with absorption peaks at 418, 529 and 541 nm (Chalazonitis, 1964; Gotow 1975; Hisano et al., 1972). In the hardshell clam *Mercenaria mercenaria,* the action spectrum showed a maximum at 510 nm, suggesting a rhodopsin (Weiderhold, MacNichol, and Bell, 1973).

In molluscs, the gastropod *Onchidium verraculatum* possess orange-pigmented neurons in the sub- and supra-oesophageal ganglia that are light-sensitive. Extraction of neural tissue yielded a red pigment that was identified as a heme-protein, and a yellow pigment identified as a carotenoid. Similar evidence for a heme-protein and a carotene-protein was obtained from the orange-pigmented neural tissue of the snail *Lymnaen stangnalis* (Benjamin and Walker, 1972). As we previously mentioned, in cephalopod molluscs the extraocular photoreceptors are the parolfactory vesicles, which are closely associated with the brain. From the parolfactory vesicles of the squid and from the episteller body on the surface of the stellate ganglia of the octopus, rhodopsins were isolated (Baumann et al., 1970; Mauro, 1977; Mauro and Baumann, 1968). In the deep-sea squid *Todarodes pacificus,* both a retinochrome and rhodopsin have been identified and isolated from the parolfactory vesicles (Hara and Hara, 1980).

Attempts to isolate and identify the extraocular photoreceptor pigment molecule for vertebrates have been more difficult. The electrophysiological and spectral sensitivity data obtained for frog skin and for the frog frontal organ are strikingly similar. The spectral peaks for both tissues lie in the ultraviolet and visible regions (355 and 515 nm for the frontal organ, and 385 and 500 nm for the frog skin). These data suggest that the pigment molecule resembles a retinal and rhodopsin-type system (Rayport and Wald, 1978). Additional evidence for a rhodopsin photosystem was obtained from spectral sensitivity measurements of the dark-adapted frog pineal, whose maximum was 502 nm (Hartwig and Baumann, 1974). Also, in the small spotted dogfish shark, *Scyliorhinus caniculus,* the pineal spectral sensitivity was around 500 nm (Hamasaki and Streck, 1971). However, in the mouse and rat, where the Harderian gland is implicated as an extraocular photoreceptor, a reddish brown pigment was isolated, and the absorption spectrum of the extracted pigment indicates a porphyrin.

The evidence from behavioral action spectra and spectral sensitivity measurements of the extraocular receptor sites point to the fact that similar pigments are involved in the photosystem. Where the pigments could be extracted from extraocular tissue receptor cells, a rhodopsin system was identified as the photoreceptor molecule. The behavioral and spectral sensitivity response to blue light suggests that flavins participate in the photoprocess, for flavins have been identified with phototactic behavior in bacteria, in fungi, in protozoans, and in insects (Ninneman, 1980; Wolken, 1975). However, other pigments, including carotenoids and porphyrins (hemes and cytochromes), are implicated as well as in these photoprocesses.

LIGHT AND HORMONAL STIMULATION BY THE PINEAL

The pineal has a long history that goes back to India, for it is believed that the Hindus, more than three hundred thousand years ago in their literature of enlightenment, recognized that the pineal was an eye. Rene Descartes (1637) thought that the pineal was the seat of the soul. He visualized that the events of the world are perceived through the eye and transmitted through a series of fibers to the pineal organ in the brain (Fig. 13.1). It was believed to be the vestigal organ referred to as the "third eye" of our reptilian ancestry. The pineal is an odd mass of cells located at the base of the brain near the very top of the spinal column (Fig. 13.2). It is a small greyish structure about 6 mm long shaped like a pine cone, from which it got its name, and weighs about 0.1 g in humans. Serious investigations of the pineal began in the 1880s, but it was not until Bargmann (1943) proposed that the pineal was regulated by light, and Wurtman and Axelrod (1965) proved this to be so, that the pineal gland began to be understood. We now know the pineal is an extraocular photoreceptor functioning as a sensitive neuroendocrine transducer, a light-activated biological clock. A large body of information has now become available about the physiological factors that control pineal activity (Arendt, 1985; Ehrenkranz, 1983; Tamarkin, Baird, and Almeida, 1985; Wurtman, 1975; Wurtman, Axelrod, and Kelly, 1968).

Several studies of the pineal deal with its behavior. It was observed that if crushed pineal glands were introduced into water in which tadpoles were swimming, the tadpole skin color bleached. Lerner and his associates (1958) were the first to see a relationship between the extract of the pineal glands and skin coloration. They isolated a substance from bovine pineal glands and called it melatonin, because it caused the contraction of the melanin granules. The action of melatonin on the skin indicates that the pineal organ

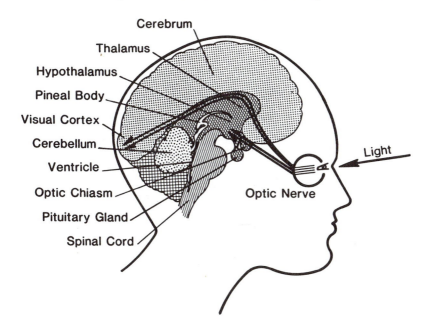

Figure 13.1 Diagram of how light through the eye passes to the brain, where the pineal gland is activated.

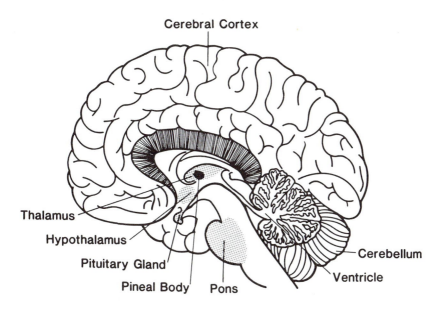

Figure 13.2 Diagrammatic of the brain's limbic system (concerned mainly with memory).

of fish and amphibians causes changes in skin pigmentation in response to light. It was then proven that melatonin was a hormone; the fact that an active substance was produced by the pineal demonstrated that it was not simply a vestigal organ. The melatonin molecule was found to be the indole amine *N*-acetyl-5-methoxy-tryptamine (Fig. 13.3). The precursor chemical necessary for melatonin synthesis was then found to be serotonin. Serotonin is relatively widespread, for it is found in the cephalopod molluscs, the amphibians, and in the pineal of all vertebrates. Surprisingly, it is also found in fruits, some of which are bananas, figs and plums.

The synthesis of melatonin begins with the amino acid 5-hydroxy-tryptophan. Enzymatic action removes the carboxyl group (COOH)). The product of this reaction is serotonin (Fig. 13.3). Another enzymatic reaction acetylates the molecule to form *N*-acetylserotonin, which is then methylated to yield melatonin. The methylating enzyme hydroxyindole-o-methyl-transferase is found only in the pineal of mammals. The activation of this pathway is controlled through a regular oscillating circadian rhythm.

In vertebrates, the concentration of serotonin in the pineal is highest during the day and lowest at night or in darkness. Just the reverse was found for melatonin, which is highest at night and lowest in continuous light. So the melatonin rhythm exhibits 24-hr periodicity, which is directly related to the circadian rhythm. The ability of melatonin to modify gonadal function suggests that its secretion has to do with the timing of the estrus and menstrual cycles of the reproductive processes.

In vertebrates the pineal has evolved active photoreceptor structures, and the level of function in the present-day mammalian pineal continues to be

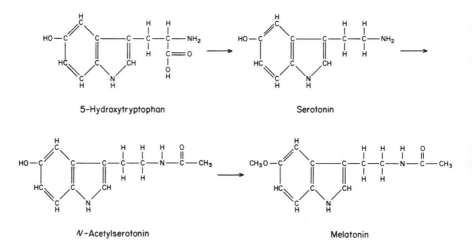

Figure 13.3   Synthesis of melatonin (Refer to Axelrod, 1974).

regulated by light. A striking finding is the similarity between the structure of the photoreceptor cells of the pineal in lizards and amphibians to that of the retinal rods and cones of vertebrate eyes (Eakin, 1965, 1973; Kelly, 1965, 1971). The action spectrum for the frog pineal has a maximum near 560 nm, which suggests that the photoreceptor pigment is a rhodopsin comparable to the cone visual pigment, iodopsin (Fig. 12.2).

An interesting finding is that serotonin production and various mental states may be related. The structures of serotonin and hallucinogenic (or psychotomimetic) agents are similar. For example mescalin, extracted from the peyote cactus, is structurally similar to noradrenalin (Figs. 13.4 and 13.5). Serotonin is known to have the same type of physiological effect as adrenalin. It is curious that certain structural features of serotonin bear a relation-

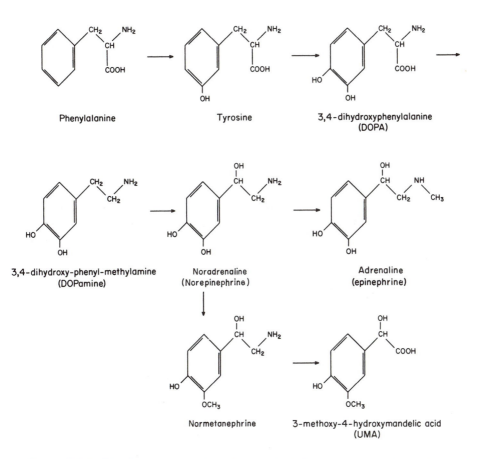

Figure 13.4 Synthesis of noradrenaline and related compounds from phenylalanine and tyrosine.

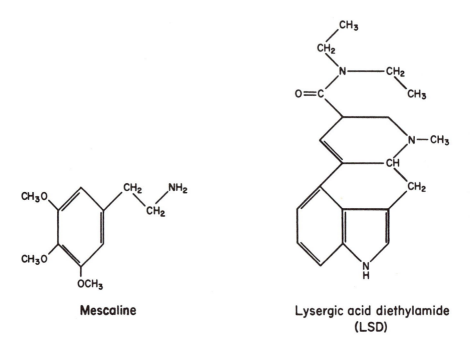

Mescaline   **Lysergic acid diethylamide**
                                              (LSD)

Figure 13.5 Structural relationships of mescaline and LSD to DOPA and noradrenaline and to the catecholamines. (Refer to Figs. 13.3 and 13.4.)

ship to lysergic acid diethylamide (LSD). These drugs then exert their psychical effects on the brain by competing with the enzymes and receptor membranes of the pineal that are involved in the metabolism and activity of serotonin.

PHOTOPERIODISM IN PLANTS

Since ancient times, it has been observed that the leaves of plants fold at night and open in the morning. It has also been observed that plants orient to varying intensities as the light changes during the day; in effect, they measure the amount of light and clock the time of day. Photoperiodism in plants, then, is the response to variations in light wavelength and the length of day and night, controlling germination, growth, and flowering. For example, continuous red light at 660 nm was fund to be effective in altering a plant's response by preventing flower formation. But it was also found that if a flash of red light was followed immediately by a short interval of far-red light with a maximum of 730 nm, the effect of the red light was cancelled. The plant then acted as if its night time had never been interrupted, and it flowered.

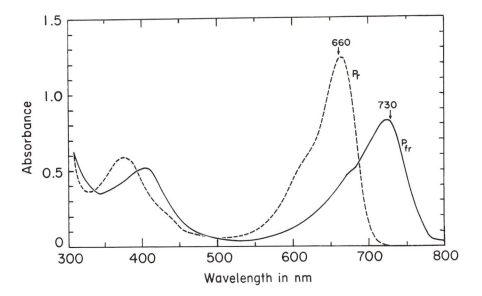

Figure 13.6 Phytochrome, absorption spectrum of the red P_r form (- - - -) and the far-red P_{fr} form (- - - -), isolated from oats. *(According to Hartmann, 1966, p. 352.)*

There appear, then, to be two light reactions, one with a maximum in the red and one with a maximum in the far-red.

The sensor pigment responsible for this red and far-red effect is the chromoprotein *phytochrome* (Hendricks, 1968). The phytochrome system allows the plant to determine day time. Phytochrome has two distinct states, one with a maximum absorption in the red near 660 nm and another with a maximum absorption in the far-red near 730 nm (Fig. 13.6).

$$P_r \underset{730 \text{ nm}}{\overset{660 \text{ nm}}{\rightleftarrows}} P_{fr} \xrightarrow{\text{darkness}} P_r$$

The structure of phytochrome (Fig. 13.7) is a linear tetrapyrolle and is structurally related to the phycocyanin pigment of the cyanobacteria (Figs. 7.6 and 7.7). The absorption spectra of phytochrome and phycocyanin also appear to be similar. Phytochrome with its system of conjugated double bonds resembles a carotenoid molecule (Fig. 7.8). In the photochemistry of phytochrome, there are similarities to the behavior of the visual pigment rhodopsin. Hendricks and Siegleman (1967) proposed two possible mechanisms for the phototransformation of phytochrome, based on the *cis* to *trans* isomerization similar to that of retinal in the rhodopsin system for vision (Figs. 12.6 and 12.14).

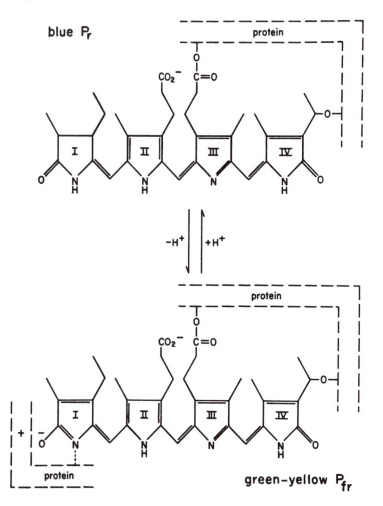

Figure 13.7   Molecular structure of phytochrome, from the red (P_r) to the far-red (P_{fr}) states. (As indicated by Rüdiger, 1969.)

Although some workers have looked for it, there is no experimental evidence so far that phytochrome is present in animals. However, in certain animals the photoperiodic effect may be due to a similarly structured photoreceptor molecule.

PHOTOPROCESSES AND LIFETIMES

We have discussed how light provides a signal to the cell, which then clocks the time. But does it also signal the end of the organism's time? Organisms from bacteria to man have fixed lifetimes; for example, the lifetime of a bacterium is in the order of minutes, protozoa and animal cells hours, and the mayfly one day, the frog six years, the dog fifteen years, and man near one hundred years. Although we may ponder the question—though in the end there may be no answer—we can try to indicate experimental directions worth considering and to speculate whether light is related to lifetime clocks.

If we put aside the genetic basis for lifetimes (those associated with the transcription and translocation processes through messenger and transfer RNA), what other factors can be considered? One suggestion is that the hormonal processes associated with the brain and nervous tissue break down. In animals this suggestion points to the brain, which regulates the release of neurotransmitter molecules, of which acetylcholine and serotonin (Figs. 9.14 and 13.3) are examples previously discussed in the control of photoperiodic phenomena.

Another direction was indicated by Strehler (1962) pointing to the fact that there is an accumulation of yellow pigments with aging and senescence. These yellow pigments are found to be associated with fluorescent granules, referred to as cellular microbodies. The microbodies appear to increase in nerve, brain, muscle, and sex glands with age, but their real function is unknown. In many cases these cell inclusions appear to be crystalline. Some of these microbodies possess enzymatic activity, such as alkaline phosphatase and acetylcholine esterase, which have been identified with lysosomes (De Duve, et al., 1955).

Attempts to isolate and extract the pigment from these microbodies have not been successful. Moreover, from extractions made from the nerve cells an absorption spectrum was obtained. The absorption spectrum has a maximum at 370 nm and its fluorescence emission spectrum was maximal in the 440–460 nm and 530–560 nm range. This pigment could be a pteridine derivative, but it more closely resembles a flavin complex and was loosely termed an "aging" pigment. Also a similar pigment found in the secretory granules was referred to as "senility" pigment (Gatenby and Leslie Ellis, 1951). Further extraction indicated it was a complex that could be partially solubilized in chloroform-methanol. The soluble fraction contained cephalin and unsaturated lipids similar to the lipids found in membranes and nervous tissue. The insoluble residue was an unidentified brown pigment, referred to as lipofucsin. Lipofucsins are now referred to as aging pigments, for during aging they accumulate in neurons, in the retinal pigment of the human eye and other tissue cells (Sohol, 1981). It is hypothesized that these lipofucsin granules are associated with the progressive decreasing rate of oxygen diffusion into these tissue cells, and hence with the aging process. The lipofucsin granules fluoresce yellow from ultraviolet excitation, and chemical analysis indicates they contain carotenoids, myoglobin and flavoproteins (Karnaukhov, 1973).

We have already encountered yellow pigments, suggesting that the flavins could be involved, for the flavins are yellow, fluoresce, and are photosensitive. We have already shown how a flavoprotein could function in photoprocesses for phototropism and phototaxis. A further finding was that the enzyme D-amino acid oxidase (FAD), whose prosthetic group is a flavin, changes with age, and is found in high concentration in nervous tissue. In addition the enzyme is found in the liver and kidney of almost all vertebrates, including birds, reptiles, and fish. It is also found in the invertebrates, the cephalopod molluscs, and the arthropods, as well as in plants, fungi and various strains of bacteria.

What is curious, though, is that a yellow pigment, neurotransmitter molecules, crystalline microbodies, and membranes seem to be somehow involved and are found to accumulate in aged animals. Light plays a role, but we know too little of these processes. However, these findings do suggest directions for future investigations.

REMARKS

A major function of extraocular photoreception is to inform the animal of the presence, direction, intensity, and wavelength of the light. The responses are observed as phototactic movements, either to or away from the light source, in movements of parts or of the whole body, and in orientation of the whole creature. Light also can set rhythmic behavior, as in circadian rhythms, in migration, and in the control of hormonal reproductive processes.

The photoreceptor sites that have been identified for dermal or diffuse photosensitivity are pigmented regions of the skin, in pigmented neurons in specific ganglion cells, and in the pineal and other cells in the brain. As to the photoreceptor molecule, the identification has come primarily from behavioral action spectra and electrophysiological spectral sensitivity measurements. Where the photoreceptor molecule has been isolated from the extraocular tissue cells, it has been identified as a retinal molecule, operating in a rhodopsinlike system, like that for vision. Even the photoreceptor structure in the pineal organ resembles the visual photoreceptor of the eye.

The effects of light on the mammalian pineal organ are mediated by a multisynaptic neuronal system involving the brain, spinal cord, and sympathetic nervous system. The pathway differs from the nervous impulses responsible for vision and is unique to mammals (Wurtman, 1975). The molecule responsible for this activation is melatonin. It is interesting that the neurotransmitter molecules acetylcholine and serotonin (the precursor molecule of melatonin) appear to play a role in photoperiodic phenomena. Neurotransmitter molecules acetylcholine, epinephrine, and norepinephrine, all of which are associated with the nervous system of vertebrates, are also found in protozoa and in invertebrate animals.

In comparing extraocular photoreception with the visual system, the threshold of light intensity necessary to produce an extraocular behavioral response is comparable to the visual threshold in invertebrates. But the threshold for vertebrate vision is much lower.

The extraocular photoreceptor system was not lost as the visual system evolved, but continued to develop via the nervous system and the brain. The extraocular receptor site in vertebrates is associated with older brain structures, the rhinencephalon and the pineal. This fact suggests a long history for the development and integration of extraocular photoreception in the vertebrate brain. The extraocular system has continued to function alone as evidenced in circadian rhythms, hormonal changes, and sexual cycles, but it can also work in conjunction with the visual system. These light-endocrine links demonstrate convincingly that the role of light in animal behavior, including human behavior, is truly extensive (Menaker, 1976, 1977; Wolken and Mogus, 1979, 1981; Yoshida, 1979; and Johnson and Hastings, 1986).

14

Light, the Brain and Memory

> The world of our senses is a world of matter and energy, space and time. After centuries of philosophical and scientific study, these the very logical elements of science, are no doubt still without a final description.
> L. J. Henderson, *The Fitness of the Environment,* 1913

> Memory is an organ to apprehend the Impression that is made in Time. By analogy, That the record in memory is like . . . Impressions of light impinged on matter . . . retain these impressions . . . though for no long time.
> Robert Hooke, *Posthumous Works of Robert Hooke,* 1705.

Living organisms receive information from their environment via their receptors. In animals the processing of this information is an essential function of the nervous system.

Light plays an important role in the behavior of living organisms; this behavior is initiated by photosensitive pigment molecules within the cell membrane or the membranes of photoreceptors. How are light stimuli processed via the receptors to cells of the nervous system and on to the brain?

The kinds of information that are important to living organisms are (1) the genetic information, which does not get feedback from the organism but is passed on from generation to generation, (2) the sensory information, which has considerable feedback into the storage system but is not passed on from generation to generation, and (3) the communicated information, which does have feedback and is passed to the next generation. It is the sensory and communicated information (memory) that concerns us here.

Learning and memory are intimately related and are two of the most complex aspects of the brain. The mammalian brain (and in particular the human brain) is a unique organ in structure and complexity compared to that of lower animals. Neuroscientists are beginning to unravel this complexity, but though much has been learned about the neurochemistry and electrophysiology of the nervous system and the brain, our knowledge of their mechanisms is far from complete, especially of the neurochemistry and neurophysiology of learning and memory.

If learning is defined as the acquisition of a response to a stimulus and memory as the storage of this response, then learning is not limited to animals with a highly developed nervous system and brain. Bacteria, fungi, protozoa, algae, and even plants react to environmental stimuli; thus, they grow and replicate, and all these processes require much information processing and storage. Nothing in their structural organization seems capable of processing and storing sensory information as effectively as the central nervous system of complex multicellular animals. Perhaps, in fact, there is no hard and fast boundary between the more organized vertebrate central nervous system and the basic single nerve cell and ganglion system, as far down the evolutionary phyla as coelenterates and annelids. For all organisms share to some degree the essential properties that underlie the versatile behavior of vertebrates: response to stimuli, growth, replication, and information processing.

The ability to learn, in terms of acquiring a response to a stimulus, goes back to primitive unicellular organisms that demonstrate simple association in response to stimuli and react via a simple change in constitution. Although incapable of long term storage, they remember this response for a limited time, a form of short-term memory. Such limited learning-memory processes no doubt serve a survival function for these organisms. On the other hand, complex learning processes associated with long-term memory are contained within the more highly developed nervous system of animals and must have evolved from simpler systems. The study of memory in such organisms may give us some insights into the memory processes in more complex vertebrate nervous systems.

Problems with hypothesis testing arise from the fact that memory is behavioral as well as physiological and is assumed to be a higher order phenomenon. As far as we know, the brain interacts with the environment through primary sensors (the visual, auditory, and tactile sensors), which detect and transduce a variety of environmental energies (light, sound, pressure) into electrochemical signals. These signals are further processed by sensors before being transmitted along established pathways through axons, neurons, and synapses to arrive completely transformed at the cerebral cortex.

To discover the physical-chemical basis of memory, two avenues of research are being vigorously explored. One is an electrophysiological basis for memory. The other is a chemical basis for memory, which is the more relevant of the two for our interests. Let us briefly review some research to discover if in fact there is a chemical basis for information processing and storage of memory.

CHEMICAL MEMORY MOLECULES

One of the earliest experiments to determine whether there was a chemical basis for transfer of memory was performed by McConnell, Jacobson, and Kimble (1959), using planarians, which have the ability to regenerate. McConnell trained planarians, after which they were cut in half transversely, and both halves were allowed to regenerate into whole planarians. The regenerated heads and tails both showed significant retention of the conditioned response learned. He concluded that the "memory molecule" could migrate and was not specifically located in the head. Additional work indicated that retention of some part of the original learning could be found in regenerated planarians with no intact structure from the trained parent. When trained planarians were fed to untrained cannabalistic planarians, some of the information acquired by the trained planarians seemed to transfer to the cannibals by ingestion (McConnell, 1962; 1966; McConnell and Shelby, 1970). These experiments suggested to McConnell that some of the memory had been coded chemically in the RNA molecule. The possible relation between memory and RNA led to the hope that learning could be transferred through "memory molecules."

The interest in RNA synthesis accompanying memory imprinting was generally related to the belief that the ultimate step in the memory process is the synthesis of specific proteins. Proteins or their aggregates were regarded as the long-lived storage records that retain imprints of experience throughout life. Hydén (1967), in search of a chemical basis for memory, implicated RNA in relating learning to memory. Hydén and Lange (1968) showed that a specific protein is associated with learning. George Ungar (1970), a proponent of the chemical basis for memory, was concerned with the role of amino acid sequences in small polypeptides and thought that some intact peptide was necessary in the processing of acquired information. He trained rats to fear the dark, and from these rat brains he obtained an extract that when injected into untrained rats caused them also to fear the dark. Ungar additionally claimed to have isolated a memory molecule from goldfish trained to avoid certain colors. The fraction isolated was a polypeptide containing 8-12 amino acid residues that he termed *scotophobin*. The amino acid residues identified in the polypeptide were aspartic acid, glutamic acid, glycine, lysine, serine, and tryosine. Ungar's experiments ruled out RNA as the memory molecule, but the possibility of whether the active peptides is combined with RNA in some complex was not eliminated.

Recently a new class of compounds that has stimulated research is the so-called neuropeptides that occur naturally in the brain and other tissues and act in a similar fashion to opiate alkaloids, such as morphine and heroin (Krieger, 1983). These studies indicate promising developments in behavioral neurochemistry and raise interesting questions as to their evolutionary origin.

A large number of peptides have been identified within the vertebrate central nervous system and in the brain, but some of these peptides are found in nonneural vertebrate tissue and in lower organisms that lack well-defined nervous systems. These peptides, though, are believed to function as neurotransmitters. In these cells and organisms they may serve as primitive elements of intercellular communications prior to development of neural and endocrine systems.

The mammalian brain does possess chemical mechanisms that could account for memory, but they are not likely to be involved in the ongoing operation of neural circuitries. Lynch and Baudry (1984) propose a calcium proteinase receptor that is triggered by some event such as a flash of light. That is, calcium rapidly and irreversibly increases the number of receptors for glutamate (a probable neuro-transmitter) in forebrain synaptic membranes by activating the proteinase enzyme. They hypothesize that this process changes the synaptic chemistry and membrane structure. They further state that this mechanism is responsible for those forms of memory localized in the telencephalon. However, a satisfactory description of the biochemical process that leads to information storage in the mammalian central nervous system and the brain has not yet emerged.

An alternative explanation is that a physical change takes place in existing molecules within the receptor membranes. Only at a later stage, when consolidation and fixation occur, do chemical processes become involved. This physical change most likely comprises conformational changes of the protein, induced by either direct contact with the stimulus or by concentration gradients of ions or small molecules. Such changes alter the permeability of membranes and release neuro-transmitter molecules such as acetylcholine and serotonin.

To further explore this idea, let us turn to photoreceptor cells that exhibit a form of short-term memory.

PHOTOCHEMICAL MEMORY

A photochemical memory is characterized by the following features: information is introduced into the memory molecule by a certain signal, for instance by a flash of light at a particular wavelength, and this information is stored (memorized) until a signal of another wavelength of light erases the information and restores the memory to its original state. The photochemical cycle is a form of short-term memory and could serve as a template for long-term memory. However, the cycle must be reversible and reproducible. When the signal is erased, the memory would be permanently lost. Used as a short-term template, this molecule could be reused once the memory is transferred.

A simple picture of the process is a reversible system that is driven by light in one direction, from A to B, and reversed from B to A in darkness or by another wavelength of light:

$$A \xleftarrow[\text{dark}]{\text{light}} B.$$

Photoreceptor molecules are light-sensitive and therefore can also be considered model systems for photochemical memory; they receive and process light information to the cell. The receptor molecules are chromophore-proteins or chromophore-enzymes that can be photosensitized by light absorption. A noteworthy example of a photoreceptor memory molecule, by our definition, is phytochrome, which functions by shifting its absorption maximum from the blue to the green-yellow. A second example is protochlorophyll, the precursor molecule for chlorophyll, which, though practically colorless, turns green when exposed to light and becomes the chlorophyll holochrome. The cytochromes can also function as memory molecules. Phytochrome, protochlorophyll, and cytochrome are all porphyrin structures, which when complexed with a protein, change color accompanied by changes in their absorption peaks. The process is reversible, depending on the wavelength of light to which they are exposed and their chemical environment.

Carotenoid pigments are additional examples of photoreceptor memory molecules. The visual pigment rhodopsin in the retinal rods is a retinal-opsin complex. Retinal is yellow, but when complexed to opsin changes color— turning red to purple in rhodopsin. Light bleaches rhodopsin to all-*trans* retinal, which turns yellow, but in the dark, retinal rearranges its structure back to the *cis* isomer and complexes with opsin again to form rhodopsin. These light-dark changes in rhodopsin are of course accompanied by shifts in absorption peaks (Fig. 12.5).

Previously we described a memory system in *Phycomyces*. If the sporangiophore receives a flash of light at a critical time in its development, and is then grown in the dark, it "remembers" where the flash of light came from and will continue to search for it in that direction. In addition, when a *Phycomyces* sporangiophore is cooled to 5°C, the sporangiophore stops growing, but a light signal can be perceived and stored so that when *Phycomyces* is returned to room temperature (22°C), the light-growth response takes place. *Euglena* also demonstrates a memory system for phototactic behavior. The photoreceptor molecule in both of these primitive organisms appears to be a flavoprotein that becomes photoreduced to a flavin semiquinone in the light and is regenerated in the dark (Wolken, 1975, 1977). Their spectral absorption peaks, in light and darkness, mimic to some extent the visual pigment rhodopsin.

Another form of photochemical memory involves the effect of ultraviolet and blue light on the genetic substrate of living organisms. Ultraviolet light brings about mutations in DNA, but blue light can repair the mechanism by a process known as photoreactivation. Photoreactivation of DNA can thus be considered another form of molecular memory that is wavelength-dependent. The process is an enzymatic photoreactivation of pyrimidine dimers. This mechanism demonstrates the organism's ability to restore itself when a memory for harmonious function has been interfered with or in effect lost.

There are, in addition, naturally-occurring pigments, such as the chromones (Benzy-γ pyrone), pyrans, and flavones (2-phenylchromone), which are particularly widespread in plants and animal tissues. Photochemical changes in these compounds can be brought about by ultraviolet light and reversed by visible light. An example is the colorless quinhydrone found in the hypobranchial gland of molluscs, which when irradiated with light turns purple. These shifts are all wavelength-dependent and can therefore be considered photochemical memory molecules.

IS THE PHOTORECEPTOR CELL A MOLECULAR COMPUTER?

The functional molecules of cells process considerable informational capacity in their molecular structure. This ability is especially true of the genetic molecules (Fig. 4.6), and it has been estimated that the genetic molecule that transfers information for the cell RNA in a bacterium would contain 10^{12} bits of bound information. A typical replicating growth rate of every 20 min would require a developmental rate of 10^9 bits/sec (Lehninger, 1965). When comparing this to the maximum likely storage capacity of the human brain, which has the order of 10^{10} nerve cells in the cerebral cortex, there would be around 10^{10} to 10^{11} bits of informational capacity. The sum sensory inflow from all the sensory organs to the nervous system has been estimated at 10^7 bits/sec; in a lifetime of 10^9 sec, the total would be 10^{16} bits (Young, 1971).

Photoreceptors receive and process information for the cell. For vision, the retinal photoreceptors of the eye transduce light into nerve impulses, signals that encode essential information about shape, pattern, and color into the the visual cortex of the brain. The informational capacity of the retinal rods and cones of the eye is 10^8 bit/sec. The human retina can then process in a lifetime of 10^9 sec 10^{17} bits of information, which is considerable.

Photoreceptors are structured of membranes. The membranes are bimolecular layers of lipids of about 100Å in thickness, with which proteins, enzymes, and photoreceptor pigment molecules are associated. There are 10^6 to 10^9 photoreceptor pigment molecules within the photoreceptor structure, arranged as monolayers on all the membrane surfaces for maximum light absorption. The cross-sectional area of each photoreceptor molecule on the membrane surface is approximately 100Å^2, and Kühn (1968) calculated that it should be possible to store a single bit on an area of $100\mathring{A}^2$. Because there are 10^6 to 10^9 photoreceptor molecules per cell, they can receive a significant amount of information for processing and storage.

Photoreceptor cells then can be considered analogs of a photochemical molecular computer. In Figure 14.1 a simple model for information reception storage and recall is illustrated. The molecular model consists of lipid membranes on which photosensitive pigment-protein molecules reside. Upon light absorption, changes in the pigment molecular geometry occur, resulting in changes in the protein to which the pigment is complexed in the receptor membrane. Light then triggers the system and the whole structure reacts like a jack-in-the-box spring. The structure can recover to its original state during darkness or when triggered by another wavelength of light. In a sense this molecular model behaves like a computer that functions through photochemical changes. Light induces a switching similar to that which occurs in silicon chips.

Modern digital computers generally do not have more than 10^8 bits of storage capacity. Therefore, the development of a molecular computer, based on the light switching of molecules, would have enormous computing ability, representing an advantage 1000-fold or more over the present silicon circuit chips, which are only two dimensional as opposed to the three-dimensional structures in photoreceptors. A quantum leap in informational capacity would result. In this context it is of interest to examine the models of Keyes, 1985 and Kuhnert, 1986.

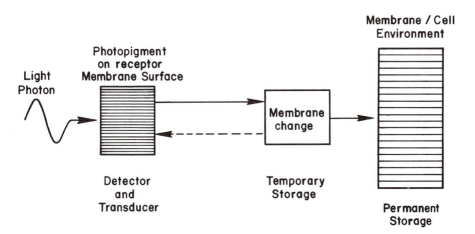

Figure 14.1 Schematic for a conformational change of the photopigment in the photoreceptor membranes upon photoexcitation.

15   *Bioluminescence*

> The production of light without heat by living things
> has always appealed to the imagination and excited the
> interest of mankind. As a remarkable example of
> functional activity in animals and plants, biolumines-
> cence itself not only presents many problems but
> has become an important means of understanding
> vital processes.
> E. Newton Harvey, *Bioluminescence,* 1952.

Let us now consider the question of how organisms emit light, for light emission is a fascinating aspect of our world of light.

The emission of light, chemiluminescence, is the result of chemical reactions. The emission of light by living organisms, bioluminescence, is therefore a form of chemiluminescence. Organisms that bioluminesce convert chemical energy into light energy with a very high efficiency, whereas in photosynthesis the energy of light is converted via chlorophyll in the chloroplasts to chemical energy. It is of interest then to see how the emission of light by living organisms came about and how it is related to photosynthesis, vision, and nerve excitation. Can we find similarities between the mechanisms of light reception and those of light emission?

Much of our knowledge of bioluminescent organisms throughout the plant and animal phyla is due to the scholarly investigation of E. Newton Harvey (1952, 1960). Organisms that bioluminesce are widely, though erratically, distributed throughout the plant and animal phyla. Of interest are certain species of bacteria, fungi, and insects that have been observed, their light described, and mechanisms studied. Numerous sea creatures bioluminesce. Among them are protozoans, dinoflagellates, hydroids, radiolarians, sponges, jelly fish, sea pens, worms, snails, clams, squid, and various fish including sharks. Among freshwater organisms, bioluminescence is rarely found, but there are some exceptions, such as the New Zealand limpet, *Latia neritoides.* No amphibian, reptiles, birds, or mammals are known to bioluminesce.

The emission of light by decaying organic matter and the emission of light by living organisms has intrigued people since it was first observed. It is recorded in ancient Chinese poetry some 3,000 years ago. Aristotle had noted that rotting meat gave off light and wrote of matter that "gives light in the dark". But it was Robert Boyle (1672) who is credited with demonstrating that rotting wood, decaying fish, and meat require air to phosphoresce. We now know that luminescence of rotting wood and plants is the result of luminous fungi. The luminescence of rotting meat and dead fish is due to the tissue being infected with bioluminescent bacteria. In most bacteria the emitted light is in the blue-green around 490 nm, but in some bacteria it ranges from 475 nm to near 520 nm. Bioluminescence of bacteria and fungi is continuous and steady but is affected by temperature changes.

One of the first to try to arrive at some chemical basis for bioluminescence was Dubois (1885). He extracted from the click beetle *Pyrophorus,* a species closely related to the firefly, two substances: one an oxidizable substrate molecule that he later named *luciferin* and another molecule, *luciferase,* an enzyme. On mixing the two together in the presence of oxygen, light was generated and he noted that no perceptible heat accompanied the light emission. This demonstration of bioluminescence is now a classical reaction:

$$\text{Luciferin (LH}_2) + \text{Luciferase (E)} \xrightarrow{O_2} \underset{\substack{\text{electronically} \\ \text{excited product}}}{P^*} \longrightarrow P + \text{light.}$$

Since these early observations, numerous investigations have been pursued to elucidate the phenomena. Our knowledge of the biochemistry of bioluminescent organisms comes mainly from the studies of Seliger and McElroy (1965), Johnson (1967), Hastings (1968), McElroy, Seliger, and White (1969) and in reviews by Cormier (1973), and McCapra (1973).

The best known example of bioluminescence is the firefly, whose flashes of light have fascinated us during the summer nights in the eastern United States. These are *Photuris pennsylvanica,* whose lantern flashes a greenish light, and *Photuris pyralis,* whose light is yellow. Fireflies inhabit every continent except Antarctica. There are more than two thousand firefly species, of which about a hundred are known to live in the United States. The most dramatic firefly displays in the world occur along certain river banks in the Orient. In most cases, the female of the species is the luminous one. Such displays, some believe, represent communal efforts to attract prey or to divert predators rather than serving as individual mating signals.

In different species of firefly, bioluminescent emission varies from about

550 nm to 595 nm. These variations have been associated with differences in the luciferase rather than the luciferin. Firefly luciferase has been isolated and crystallized and the structure of luciferin is also known. The luminescent reaction requires the enzyme luciferase to catalyze the reaction between luciferin and ATP, forming a luciferin-AMP compound and releasing pyrophosphate plus light. The overall reaction would be:

$$LH_2 + ATP + O_2 \longrightarrow L = O + AMP + PP + H_2O + light.$$

At present, investigators are inclined to the theory that through the eye, the individual firefly's central nervous system judges the timing of the discharges. The important thing to note here is that the light emission is under nervous control and that the chemistry of the luminescent reactions may be associated with a nerve impulse through the neurotransmitter molecule acetylcholine.

Bioluminescence is more prevalent in marine than in terrestrial animals. Surface and shallow water contain organisms that bioluminesce. Beebe (1935) and other oceanographers in the 1930s descended several thousand feet in bathyspheres. These explorers discovered that not only was there abundant life within certain deep layers of the ocean, but that below 300 m nearly all living creatures were bioluminescent. It is within such an environment that evolutionary forces must have been highly selective for the development of bioluminescent organisms. The relationship between depth in the sea, bioluminescence, and the bioluminescent organisms has been studied by Clarke and Denton (1962) and Denton, Gilpen-Brown, and Wright (1970, 1972). Most marine bioluminescence is predominantly blue, but organisms are found that emit green, orange, and red light as well.

Light is produced by animals with and without eyes, by sessile as well as motile animals. The bioluminescence occurs only upon stimulation and the excitation appears to be associated with a nerve impulse. Certain marine organisms, including fish, have luminescent organs whose light is sufficient to excite the eye of these animals, and in some cases luminescent bacteria are located in, or in a pouch under, the eye of these animals, as observed in the flashlight fish. Bioluminescence for these animals is a means to attract prey, to distract predators, to communicate with one another, and for sexual mating.

Let us examine several different kinds of organisms with regard to their bioluminescent mechanisms and structures for light emission. In the coelenterates, the jellyfish *Aequorea aequorea* possesses blue luminescent photophores around the edges of the mantle. Although it contains a luciferin-luciferase system, oxygen is not necessary. The system does require the addition of a calcium ion for full luminescence. The sea pansy *Renilla reniformis* is a pinkish violet colonial coelenterate that lives on the sea floor at depths of 6 m to 15 m. When a sea pansy is stimulated mechanically or electrically, it sends a concentric wave of blue-green light from the point of stimulation across its body surface. The wave of luminescence is controlled by some kind of nervous network. Aquorin isolated from these jellyfish (in Puget Sound) gives off a blue bioluminescence, but when excited by ultraviolet light it fluoresces green.

Among the crustacea is the crustacean *Cypridina hilgendorfii* found mainly in the coastal waters of Japan and also off Jamaica. This small crustacean is about 2 mm by 3 mm. It possesses a gland near the mouth that, when it is disturbed, squirts luciferin and luciferase into the sea, where they mix and react with oxygen (one oxygen molecule per luciferin molecule) to give a flash of blue light around 465 nm. In the shrimp, *Meganyctiphanes norvegica,* the light emitted from its eye is in the form of coherent light, like that from a laser (Bassot, 1966). If so, this animal evolved an optical device long before man's ingenuity ever developed the laser.

The black dragonfish, *Idiacanthus antrostomus,* emits light in the red end of the spectrum. Most sea animals accustomed to the blue-green light that penetrates the ocean are blind to red wavelengths, for their retinal visual pigments are insensitive to this part of the spectrum. But the dragonfish evolved a visual pigment to see in the red, permitting the fish to stalk its prey in the dark—a snooperscope to see but not be seen.

Bioluminescence is another facet of polychaete biology where functional explanations are incomplete, although this phenomenon occurs fairly commonly among certain families. Light flashes in the dorsal scales (elytra) of polynoids, which are controlled indirectly by the nervous system, have been regarded as a sacrifice lure and may, therefore, have a protective value. Only within the genus *Odontosyllis* (family Syllidae), is there a clear functional correlate between light emission and vision. In the waters of Bermuda, one of the interesting natural phenomena to observe is the spawning of the polychaete annelid, *Odontosyllis enopla,* accompanied by brilliant bioluminescent displays. It is believed that Christopher Columbus recorded this observation in his diary, and the phenomenon has been observed by numerous tourists to the island ever since. It takes place shortly after the full moon when the female leaves her coral burrow and rises to the surface in shallow water bays and channels, and bioluminesces. The males are visually attracted to the luminescent females near the surface and give off brief flashes of light themselves as they circle the female during spawning. In the process, the female releases her eggs in a luminous secretion. These spawning sequences occur monthly, and considerable interest has been centered around the precise lunar periodicity of this phenomenon. Such behavior begins a day or two after the full moon and peak activity usually occurs on the third evening. Bioluminescence and spawning occur periodically around 50-55 min after sunset and last for about 20-30 min.

Odontosyllis' bioluminescence is maximal in the green portion of the visible spectrum, λ_{max} from 504 nm to 507 nm. The eyes of *Odontosyllis* exhibit maximum sensitivity to green light as well, λ_{max} from around 510 nm to 520 nm, as determined by electroretinogram recordings made in response to light stimuli at various wavelengths (Wilkens and Wolken, 1981). Spectral sensitivity in the eyes of *Odontosyllis* is, therefore, closely matched to the spectral properties of the bioluminescent mating signal (Fig. 15.1). The emission spectrum of the in vitro luciferin-luciferase bioluminescent reaction of *Odontosyllis* peaks at 507 nm (Shimomura, Johnson, and Saiga, 1963) and a 504 nm maximum for the emission spectrum of *Odontosyllis* has been obtained for oxy-luciferin, the in vivo light emitter (Trainor, 1979).

Bioluminescence with an emission maximum in the green is also characteristic of a number of polychaetes that, like *Odontosyllis,* occupy intertidal habitats. For example, bioluminescence in *O. polycera* has been described as light green, and in a number of scaleworms (polynoids) light emission has maximum energy in the green, λ_{max} around 510-520 nm. In a number of pelagic organisms, the emission spectra have fairly broad bands extending from the blue-green to the yellow, 500-600 nm, whereas luminescence in a deep water alciopid is reportedly blue, characteristic of most luminescent

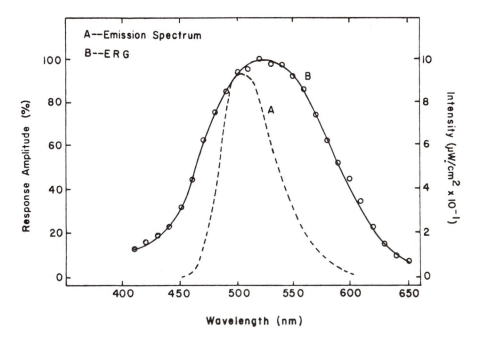

Figure 15.1 The bioluminescent emission spectrum compared to the electroretinogram (ERG) of the eye of *Odontosyllis enopla.*

marine invertebrates. What is of interest is that the wavelength of the emitted bioluminescent light matches the absorption peak for the eyes of these animals, suggesting that the bioluminescence is directly related to their mating behavior.

PHOTOPHORES

The special organs from which the bioluminescent light is generated are named *photophores*. In the dinoflagellate *Gonyaulax*, rhombohedral crystals have been identified as the photophore. These crystals were named *scintillons* and luminesce in the presence of oxygen (Hastings, 1968). Crystalline bodies have also been observed in other organisms and identified as photophores. There are, of course, many different structural types of photophores found among the various luminescent animal species.

Biologists have long observed similarities in cellular structure between the eye and the bioluminescent organs. In deep sea fish and in both squid and shrimp, the structure is like an ocellus, which consists of a luminous layer surrounded by a cup of pigment cells (Fig. 15.2). Other photophores resem-

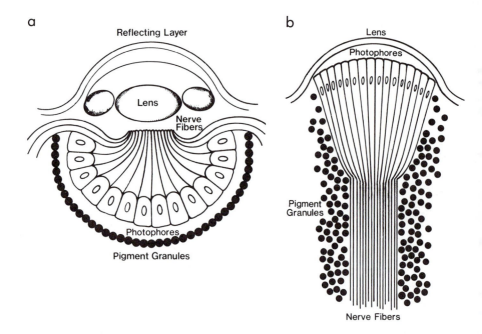

Figure 15.2   Schematized photophore of the decapod *Acathephyra (a)* and the light-producing gland cell of the shrimp *(b)*.

ble an eye with a corneal lens and a layer of luminous cells, like the retinal photoreceptors together with nerve connections.

Another interesting relationship to the retinal photoreceptors is that the luciferin-luciferase system is associated with specific membranes. In certain fish and marine worms, electron microscopy showed that photophores have lamellae that compartmentalize the biochemical luciferin-luciferase system into separate membranes (Bassot, 1966). In the Japanese squid, *Watasenia,* Okada (1966) observed membranes (microvilli) with a structure similar to that of rhabdomeres of arthropod eyes. At the level of the membrane there could be another relationship between the retinal rod and the photophore. That is, during light absorption by the retinal rod, calcium is released, and this release is believed to be associated with the generation of a nerve impulse. Similarly, during light emission in the sea pansy, *Renilla,* the nerve impulse releases calcium and light is generated.

In the firefly, *Photuris pennsylvanica,* the bioluminescent lantern structure was investigated by Smith (1963) and Wolken (1975) using electron microscopy. In our studies an interesting structural relationship was found between the insect eye and its retinal photoreceptors. In the eyes of the firefly and June beetle are tracheoles surrounding the retinal rhabdoms (Figs. 10.9, 15.3 and 15.4a). According to Miller and Bernard (1968), the ridges of the tracheoles function as a quarter-wave interference filter. This function is possible because of their twisted lamellar structure, and the thickness of spacing of the lamellae correlate with the reflected colors. The tracheoles surrounding the rhabdoms of moths are responsible for the eye glow; in butterflies this glow is seen as red. Tracheoles are also found in photocytes, the luminescent cells in the firefly lantern (Fig. 15.4b).

EVOLUTIONARY CONSIDERATIONS

There appears to be no discernible pattern of bioluminescence among the various organisms studied. But the widespread occurrence of bioluminescent organisms in the sea may suggest an important key to uncovering its origin and evolution. Bioluminescence must have had some evolutionary significance and performed a functional role that is seen in organisms that still possess it. Bioluminescence probably developed in a sporadic fashion throughout the course of evolution. If evolution at the molecular level proceeds by small changes in preexisting molecules, it may well be that luciferins and luciferases evolved from molecules that once had entirely different functions. There is some belief that the origin of bioluminescence is an evolutionary relic from the time when there was a sudden appearance of oxygen in the atmosphere, requiring a biochemical means for the removal of that oxygen. Hence the reactions were detoxifying processes necessary for the survival of these early anaerobic bacteria.

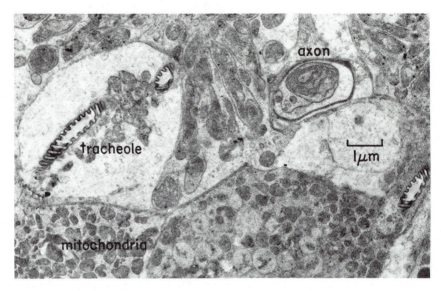

a

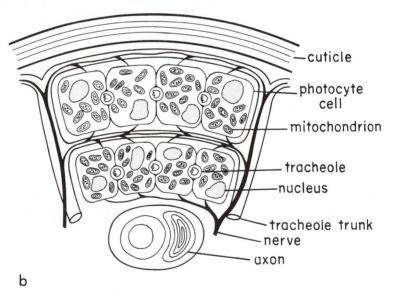

b

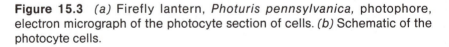

Figure 15.3 *(a)* Firefly lantern, *Photuris pennsylvanica,* photophore, electron micrograph of the photocyte section of cells. *(b)* Schematic of the photocyte cells.

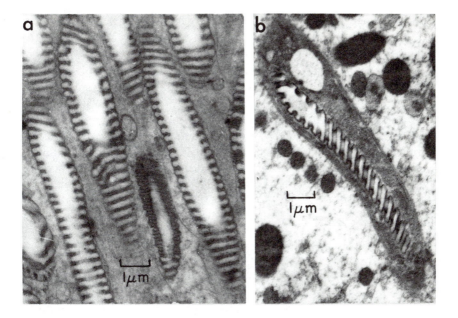

Figure 15.4 Tracheole structure, longitudinal sections in *(a)* eye of June beetle (Scarab, *Phyllophaga)* in rhabdom photoreceptor area; *(b)* in the lantern of the firefly *(photuris pennsylvanica)* in photocyte cells. *(Refer to Wolken, 1975, p. 271.)*

In bacteria bioluminescence may have an important functional biochemical role, for present day anaerobes, which are still subject to oxygen toxicity in the environment, have not been found to utilize bioluminescence as a detoxification mechanism. Hastings (1968) suggested that the bioluminescent reactions result in the formation of an active oxygen molecule, which then acts in the cell as a high energy oxidant. In this regard, the depths of the ocean probably represent an ancient and constant environment. Here the very low light level may have forced adaptation of a special mechanism by an organism to produce active oxygen. The active oxygen then could provide a chemical mechanism to allow cells to metabolize compounds not readily attacked by its enzymes.

All bioluminescent species use a similar biochemical system, that is, a luciferin and a luciferase reaction as the basic reactants that result in light emission. In addition to the luciferin and luciferin system, flavins are associated in the luminescent reaction of the firefly, the luminous exudate of earthworms, and characteristics of other organisms. We have also mentioned that flavins are involved in phototropic and phototactic behavior. There are several other points worth mentioning here. The reactions that result in bioluminescence could well be very close to the reversal of photo-synthesis. These two processes, photosynthesis and bioluminescence are found in the same organism. Bioluminescence of the dinoflagellate *Gonyaulax* can be stimulated by light in the red region of the spectrum at 675 nm (the absorption peak for chlorophyll), and the emission spectrum is at 478 nm, which is identified with luciferin (Sweeney, Fork, and Satoh, 1983).

A number of theories have been proposed to explain the significance of bioluminescence, most of which consider light emission as a form of inter-specific communication used primarily for protection. The use of biolumi-nescence as a food-securing mechanism is not certain, for many bioluminescent organisms are functionally blind. From a behavioral standpoint biolumines-cence appears to be significant, for it is a means of recognition to attract prey, to distract predators, to communicate with one another, and for sexual mating—all of which are necessary for survival.

It would be extremely interesting from an evolutionary point of view if the development of the photophore for light emission and the photoreceptors of the eye for light absorption have similar origins. Bioluminescence in animals is believed to have evolved after the development of the eye and the visual system, but whether or not the photophores for bioluminescence and the photoreceptors for visual excitation have similar origins needs to be investi-gated. There is a definite relationship between the photophores to vision in oceanic animals (Denton et al., 1985). Clearly there is an evolutionary relationship between light emission and light reception at the biochemical level that served to increase the efficiency of both of these energy processes.

16

The sun and sunlight have been associated since the earliest records of human existence with the power that its light exerts on life. Within this book we have tried to establish the scientific basis for these beliefs and to show how intimately life on Earth is related to the energies of solar radiation.

The electromagnetic spectrum of energy that reaches the earth from the sun ranges from near 300 nm to beyond 900 nm, from the ultraviolet to the infrared (Fig. 2.3). Ultraviolet radiation below 300 nm is cut off, absorbed by ozone in the upper atmosphere. Radiation beyond 900 nm, the infrared, is mostly absorbed by atmospheric water vapor and water.

Therefore, the limits of radiation effective for photobiological phenomena are less than those for photochemistry and are restricted to the visible. Why these wavelengths of light? One reason is that these are the energies used in chemical reactions and another is that these wavelengths of light are the ones used in photochemical reactions due to the molecular structure of the photoreceptor molecules in living organisms. Thus, it is no accident that life on Earth has been remarkably efficient in its use of this range of energy.

As living organisms evolved they adapted to this narrower region of the visible electromagnetic spectrum, from about 340 nm to 760 nm, with a maximum in the blue-green around 500 nm, about which most photobiological phenomena cluster (Fig. 2.4). We observed this range in examining phototropism, phototaxis, photosynthesis, photoperiodism, and vision.

We emphasized the importance of near-ultraviolet light from 300 nm to 400 nm and blue light from 400 nm to 500 nm, both of which have pronounced effects on living organisms. Among these effects are phototactic movement, timing mechanisms, and biological clocks for both plants and animals. The spectral sensitivity for vision of most insects shows a major response near 360 nm, and visual response in animals is best around 500 nm to beyond 600 nm. Radiation in the red part of the spectrum from 600 nm to 700 nm is utilized for chlorophyll synthesis and photosynthesis. Radiation from 660 nm and into the near-infrared is important for plant and animal growth, the timing of plant flowering, sexual cycles in animals, and pigment migration. Photoperiodism in plants is controlled in the near-red part of the spectrum by the shifting of light between 660 nm and 730 nm. Bacterial photosynthesis extends even further into the red, to 900 nm.

In all photobiological processes photoreceptor molecules are necessary to absorb the energy. The photoreceptor molecules are the carotenoids, chlorophyll, flavins, phytochrome, retinal, and other related pigment molecules. These pigments do not function alone in a cell but are complexed with a protein that is associated with membranes for maximum light absorption. The photoreceptors provide the means for plants and animals to detect and measure the kind of light in their environment.

The photoreceptors in these photoprocesses receive and transduce the light energy to other energies, for example, light energy $\rightarrow$ mechanical energy $\rightarrow$ chemical energy $\rightarrow$ electrical energy. In these processes light serves two major functions for the organism: one is the direct conversion of the light energy to chemical energy as in photosynthesis, and the other is the transmission of information through the sensory system, as in the visual system. Scientists are just beginning to understand the underlying mechanisms of these photoprocesses.

PHOTOSYNTHESIS

The direct conversion of light energy to chemical energy is carried out by photosynthetic bacteria, algae, and plants. This photochemical step is of major importance to all animal life on our planet. Not only does it provide a renewable source of food, but in the process of photosynthesis the oxygen concentration in our atmosphere is maintained near the 20% necessary for the respiration of animals.

To bring about photosynthesis two reactions are necessary: one is the light reaction, which is the photolysis of water, and the other is the dark reaction, which is the reduction of carbon dioxide to carbohydrates. The initial photophysical event is the absorption of light by chlorophyll *a* in the chloroplast, resulting in an excited chlorophyll molecule in which an electron is raised from its normal energy level to a higher energy level. Such excited electrons are transferred from chlorophyll to the enzymes, ferredoxin, cytochromes, to flavins and quinones. During this cyclic flow of electrons, the energy that the electrons initially acquired is transferred through oxidation-reduction reactions. These reactions are driven by two photochemical pigment systems, Photosystem I and Photosystem II (Fig. 8.4). The reduced electron acceptor of Photosystem I reduces $NADP^+$ via ferredoxin. The oxidized electron donors of Photosystem II produce oxygen by the oxidation of water. These two photopigment systems provide the high energy phosphates (reduced NADP and ATP) needed for the synthesis of carbohydrates and proteins from CO_2 and water.

Research to duplicate the process of photosynthesis with chloroplasts outside of plant cells and with synthetic systems to recover hydrogen and oxygen, which we can use as energy sources for our technology, are progressing. Although much progress has been made in demonstrating the feasibility of electron and energy transfer in such systems, we have fallen far short of the efficiency of the chloroplast in photosynthesis. Other research is directed to engineering the genes of the chloroplast, through molecular genetics, to produce plant strains resistant to environmental extremes, disease, and insects, thus greatly improving crop yield for feeding the increasing world population.

VISION

The function of light in vision is for photoexcitation of the visual pigment rhodopsin in the retinal photoreceptors, the rods and cones of the eye. In the photochemistry of the retinal photoreceptors, light energy is transduced to chemical energy, and then to electrical signals that are transferred via the optic nerve to the visual cortex of the brain.

In early invertebrate animals photosensory cells gave rise to the first eyes and primitive vision. Various kinds of optics evolved for imaging, from simple pinhole to compound and to refracting-type eyes of vertebrates (Fig. 10.1). The retinal photoreceptors were formed of membranous processes of the receptor cells. The visual pigment rhodopsin resided on or within the membranes of the photoreceptors. The photoreceptors were highly ordered structures like a crystal and were strikingly similar to the structure of chloroplasts (Figs. 8.5 and 8.6).

At the molecular level, not only do we find similarities in the structure for all photoreceptors, but for visual excitation all photoreceptors depend for their function on a single molecular group of pigments, the carotenoids. An important change occurred during the evolutionary development of animals that made them dependent on the ingestion of plants for their source of carotenoids. It was not the ingested plant carotenoids themselves but their degraded derivative, vitamin A, that became necessary for animal life. The carotenoids, then, play a central role in the biochemical evolution from the plant C_{40} (β-carotene) $\rightarrow$ to animal C_{20} (vitamin A) $\rightarrow$ retinal (vitamin A aldehyde), in the visual pigment rhodopsin.

All visual pigments, from both invertebrate and vertebrate eyes, are rhodopsins that contain the chromophore retinal$_1$ or retinal$_2$ (Fig. 12.4). Rhodopsins in the retinal rods and cones are retinal-protein complexes, and it is the 11-*cis* geometric isomer of retinal that is the functional molecule in rhodopsin. The only action of light in visual excitation is then to transform 11-*cis* to all-*trans* retinal (Figs. 12.6 and 12.14). Opsin, the visual protein, is species-specific and determines the spectral sensitivity, the absorption peak of the rhodopsin in accordance with an animal's habitat.

The visual chromophore retinal is found not only in animals with eyes but also in the cell membrane of halophilic bacteria, *Halobacterium halobium*. The bacteriorhodopsin in the cell membrane is similar to animal rhodopsin, but in these bacteria rhodopsin functions in photophosphorylation similar to the action of chlorophyll in photosynthesis. There are also indications that a retinal complex, a rhodopsin, is found in the fungus *Phycomyces* for phototropism and is the photoreceptor molecule for phototaxis in the algal flagellate *Chlamydomonas*.

The fact that a retinal complex, rhodopsin, is found in bacteria and algal flagellates, which have no eyes or nervous systems, suggests that retinal has a more generalized biological function than simply for vision. Perhaps this situation resulted from the evolution of a molecule that was synthesized early in the history of life and served other functions. But, once retinal became incorporated into the eye, it alone was used for all visual systems that later evolved.

How we see involves more than just what takes place in the visual pigments of the retinal rods and cones. The mechanism includes not only the absorption of light by the visual pigment but also an elaborate transduction process. Many questions remain as to how a single photon of light leads to visual excitation, exactly how the retinal photoreceptors transduce light energy to electrical signals, and how all this information is transferred via the optic nerve to the visual cortex of the brain where the images of our world are recorded.

PHOTOPERIODISM AND EXTRAOCULAR PHOTOSENSITIVITY

Light and dark periods control life rhythms. Light can set the rhythmicity of an organism's behavior—its biological clock. Photoperiodism is seen in plants in leaf movements, in the control of flowering cycles, and in animals in movement and navigation, and in sexual reproductive cycles.

The photoperiodic behavior of higher plants is due to the action of the phytochrome molecule. The structure of phytochrome is an open ring porphyrin with which its system of conjugated double bonds resembles a carotenoid molecule. The phytochrome transformation, from the red-absorbing to the far-red-absorbing form, involves a change in the molecular structure of phytochrome. In the photochemistry of phytochrome there are similarities to the visual pigment rhodopsin, as both are based on a *cis* to *trans* isomerization and are sensitive to the dark/light transitions.

In animals, eyes are not the sole means of photoreception, for the general body surface is remarkably light sensitive. This diffuse photosensitivity over the whole or part of the animal's body, the dermal light sense, is especially keen in eyeless and blinded animals. Even deeper tissue cells in the body are photosensitive, including nerve and brain cells. Such extraocular photoreception informs the animal of the presence, direction, intensity, and wavelength of the light. Hence animals possess a photoreceptor system to measure all aspects of the light stimulus. The responses are observed as phototactic movements, either to or away from the light source, of parts of or of the whole body, and in orientation of the animal. Extraocular photoreception can also set rhythmic behavior, as in circadian rhythms in the control of hormonal reproductive processes, and even in migration.

The photoreceptor sites that have been identified for extraocular photosensitivity are pigmented regions of the skin, pigmented neurons in specific ganglion cells, and the pineal organ in the brain. At these sites, where a photoreceptor has been located we see a structure surprisingly similar to the retinal rods and cones of the eye. There is experimental evidence that the pigment molecule is a retinal complex, as is the visual pigment rhodopsin, but other pigments have also been implicated.

The mammalian pineal is regulated by light, and the behavioral action spectrum indicates a maximum peak near 560 nm, similar to the vertebrate retinal cone pigments. But the subsequent pathway in this process differs from the nervous impulses responsible for vision and is unique to mammals. The molecule responsible for this activation is melatonin, which in bright light decreases and in darkness increases in concentration. It is interesting that the neurotransmitter molecules acetylcholine and serotonin (the precursor molecule of melatonin) also play a role in photoperiodic behavior. The pineal then serves as a light-activated neuro-endocrine transducer that functions for the animal as a biological clock.

In comparing extraocular photoreception with the visual system, the threshold of light intensity necessary to produce a behavioral response is much lower than for vertebrate vision, but is comparable to the visual threshold in invertebrates. As the visual system evolved, the extraocular photoreceptor system continued to develop via the nervous system and the brain. The extraocular receptor site in invertebrates is associated with older brain structures, that is, the rhinencephalon and the pineal, suggesting a long history for the development and integration of extraocular photoreception in the vertebrate brain. The extraocular system has continued to function alone as evidenced in circadian rhythms, hormonal changes, and sexual cycles, but it also functions in conjunction with the visual system. These light-endocrine links demonstrate that the role of light in the behavior of animals, including human behavior, is truly extensive (Menaker, 1976, 1977; Wolken and Mogus, 1981; and Yoshida, 1979).

LIGHT AND HEALTH

There are many beneficial effects of sunlight on the body, from warmth to recovery from states of depression and to feelings of well-being. The therapeutic use of light has been recognized for some time, but we are just beginning to explore a wider range of applications for combating various diseases that afflict us and to our health in general.

One of the early discoveries about light for health dealt with the importance of vitamin D to bones. Vitamin D is formed naturally upon the absorption of near-ultraviolet radiation by its inactive precursor, the sterol 7-dehydro-cholesterol, in the skin of amphibia and reptiles, on the feathers of birds, on the hair of animals, and inside human epidermal cells. It produces the active substance cholecalciferol or vitamin D. The lack of vitamin D in humans results in the inability of the body to deposit calcium and phosphorus in bones. Before vitamin D therapy, destructive bone diseases were found to occur in northern climates where there is little sunlight, but were unknown to inhabitants of southern regions. It was then found that the amount of sunlight is directly related to calcium metabolism and its increased absorption in the bone.

Constant exposure to the sun produces sunburn, the result of ultraviolet radiation on the human skin. Light penetrates through the skin from the extent of 10% to as much as 50% and the melanin pigments filter out much of the ultraviolet effect. However, chronic sunburn is injurious to the skin; it results in premature aging, and prolonged sunburn can lead to skin cancer. The action spectrum for skin cancer is in the range 260-300 nm that brings about alterations of the DNA molecule. Ultraviolet radiation at 260 nm is the absorption peak of DNA and 280 nm is that for proteins; therefore absorption in this range of radiation produces damaging effects to cells and greatly increases the frequency of cellular mutations.

Early in the history of life when there was much more ultraviolet radiation than at present, the synthesis of retinal by living organisms may have been to protect DNA against radiation damage or to promote DNA repair. In this context it is important that vitamin A, β-carotene and retinal may have a protective effect against the onset of certain types of cancer; this possibility is being clinically tested. Vitamin D, essential for a healthy skeleton, may have also been one of the molecules that evolved to serve as an early protecting sunscreen against ultraviolet radiation. Surprisingly, the damage resulting from ultraviolet radiation can be reversed by visible light, specifically blue light from 430 nm to near 500 nm, by a phenomenon known as photoreactivation. This repair mechanism is effective not only for bacteria, algae, fungi, and plants, but also for animals, including human cells.

There are other damaging effects of near-ultraviolet radiation and visible light. They are brought about by a photosensitizing molecule and molecular oxygen. Photosensitization is produced during the absorption of light by a molecule that then becomes activated and causes destructive photo-oxidation to the cell. This photobiological phenomenon is known as photodynamic action. Photosensitization affects many types of cells (including the skin) that absorb light and thereby become sensitized. This sensitized state can be produced by naturally-occurring chemical photosensitizers in the cell, or by certain medications, drugs, and pollutants in the air. But advantage can be taken of this mechanism by using light-activated drugs to selectively target certain types of cancer cells.

Light therapy is being used to treat autoimmune system diseases. The systematic changes in the immune system are an important factor in association with ultraviolet radiation with malignancy. And near-ultraviolet blue light dramatically affects the treatment of jaundice in newborn infants (Lightner and McDonagh, 1984). We are just beginning to explore and understand the medical uses of light to treat a variety of diseases and health-related problems (Morison, 1984; Wurtman, Baum, and Potts, 1985).

Light serves two major functions for plants and animals. One is the direct conversion of light energy to chemical energy for photosynthesis, and the other is transmission of information about the environment, resulting in movement, orientation, and vision. Photoreceptor systems also affect growth, hormonal stimulation, clocking mechanisms, and sexual cycles. The photoreceptors receive, and in the process transduce, the energies of light to other forms of energy as light energy $\rightarrow$ mechanical energy $\rightarrow$ chemical energy $\rightarrow$ electrical energy.

Light is truly a life force, as essential millions of years ago for the first development of life as it is today for the continuing survival of all life on Earth.

References

Abrahamson, E. W., and R. S. Fager, 1973, The chemistry of vertebrate and invertebrate visual photoreceptors, *Curr. Top. Bioeng.,* pp. 125-200.

Amesz, J., 1973, The function of plastoquinone in photosynthetic electron transport, *Biochim. Biophys. Acta* **301:**35.

Andrews, E. A., 1892, On: The eyes of polychaetous annelids, *J. Morphol.* **7:**169-222.

Arden, G. B., C. D. B. Bridges, H. Ikeda, and I. M. Siegel, 1966, Rapid light-induced potentials common to plant and animal tissues, *Nature* **212:**1235.

Arendt, J., 1985, The pineal: a gland that measures time?, *New Scientist* **107:**36-38.

Arnon, D. I., 1965, Ferredoxin and photosynthesis, *Science* **149:**1460.

Astbury, W. T., 1933, *Fundamentals of Fibre Structure,* Oxford University Press, Oxford.

Austin, G., H. Yai, and M. Sato, 1967, Calcium ion effects on *Aplysia* membrane ion potentials, in *Invertebrate Nervous System,* C. A. G. Wiersma, ed., University of Chicago Press, Chicago, pp. 39-53.

Autrum, H., and D. Burkhardt, 1961, Spectral sensitivity of single visual cells, *Nature* **190:**639.

Autrum, H., and V. von Zwehl, 1962, Zur Spektralen Empfindlichkeit einzelner Sehzellen der Drone *Apis mellifica, Z. Vergl. Physiol.* **46:**8.

Autrum, H., and V. von Zwehl, 1964, Die spektrale Empfindlichkeit einzelner Sehzellen des Bienenauges, *Z. Vgl. Physiol.* **48:**357.

Axelrod, J., 1974, The pineal gland; a neurochemical transducer, *Science* **184**:1341.

Bach-Y-Rita, P., and C. C. Collins, eds., 1972, *Brain Mechanisms in Sensory Substitution,* Academic Press, New York.

Bargmann, W., 1943, *Handbuch der Microskopisehen Anatomie des Menschen,* W. Mollendorf, ed., vol. 6, Springer, Berlin, pp. 309-502.

Bassot, J. M., 1966, On the comparative morphology of some luminous organs, in *Bioluminescence in Progress,* F. H. Johnson and Y. Haneda, eds., Princeton University Press, Princeton, N. J., pp. 557-610.

Baumann, F. A., A. Mauro, R. Milecchia, S. Nightingale, and J. Z. Young, 1970, The extraocular receptors of squids *Todarodes* and *Illex, Brain Res.* **21**:275-279.

Beebe, W., 1935, *Half-Mile Down,* Harcourt, Brace and Co., New York.

Benjamin, P. R., and J. S. Walker, 1972, Two pigments in the brain of a freshwater pulmonate snail, *Comp. Biochem. Physiol.* **41B**:813-821.

Berkner, L. V., and L. C. Marshall, 1964, The history of oxygenic concentration in the earth's atmosphere. *Faraday Soc. Discussions* **37**:122.

Bernal, J. D., 1933, Liquid crystals and anisotropic melts, *Faraday Soc. Trans.* **29**:1082.

Bernal, J. D., 1951, *The Physical Basis of Life,* Routledge & Kegan Paul, London.

Bernal, J. D., 1967, *The Origin of Life,* Wiedenfeld and Nicholson, London.

Bernard, C., 1866, *Lecons sur les Proprietes des Tissus Vivants,* Bailliere et Fils, Paris.

Blaurock, A. E., and W. Stoeckenius, 1971, Structure of the purple membrane, *Nature* **233**:152.

Block, G. D., and M. E. Lickey, 1973, Extraocular photoreceptors and oscillators can control the circadian rhythm of behavioral activity in *Aplysia, J. Comp. Physiol.* **86**:367-374.

Böll, F., 1876, Zur Anatomie and Physiologie der Retina, Monatsber, Preuss., Akad. Wiss., Berlin **41,** pp. 738-788.

Bowness, J. M., and J. J. Wolken, 1959, A light sensitive yellow pigment from the housefly, *J. Gen. Physiol.* **42**:779.

Boyle, R., 1672, Some observations about shining flesh, both of veal and pullet, and that without any sensible putrefaction in those bodies, Roy. Soc. London Phil. Trans. **89**:5108.

Brawerman, G., and J. M Eisenstadt, 1964, DNA from the chloroplasts of *Euglena gracilis, Biochim. Biophys. Acta* **91**:477.

Brown, A. M., P. S. Baur, Jr., and F. H. Tully, Jr., 1975, Phototransduction in *Aplysia* neurons: Calcium release from pigmented granules is essential, *Science* **188**:157-160.

Brown, G. H., 1977, Structure and properties of the liquid crystalline states of matter, *Colloid Interface Sci.* **58**:534.

Brown, G. H., and J. J. Wolken, 1979, *Liquid Crystals and Biological Structures,* Academic Press, New York.

Brown, P. K., and G. Wald, 1964, Visual pigments in single rods and cones of the human retina, *Science* **144**:45.

Bungenberg de Jong, H. G., 1936, *La Coacervation,* Hermann, Paris.

Bünning, E., 1958, Cellular clocks, *Nature* **181**:1169.

Bünning, E., 1973, *The Physiological Clock,* 3rd ed., Springer-Verlag, New York.

Bünning, I., 1986, Evolution der circadianen Rhythmik und ihrer Nutzug zur Tageslängenmessung, *Naturwissenschaften* **73**:70-77.

Burkhardt, D., 1962, Spectral sensitivity and other response characteristics of single visual cells in the arthropod eye, in *Biological Receptor Mechanisms,* J. W. L. Beaument, ed., Academic Press, New York, pp. 86-109.

Butenandt, A., V. Schiedt, and E. Bickert, 1954, Uber Ommochrome. III. Mitteilung Synthese des Xanthommatins, *Justis Liebigs Ann. Chem.* **588:**106.

Cairns-Smith, A.G., 1971, *The Life Puzzle,* Oliver & Boyd, Edinburgh.

Cairns-Smith, A.G., 1982, *Genetic Takeover and the Mineral Origins of Life,* Cambridge University Press, Cambridge.

Cajal, S. Ramón y., 1918, Observaciones sobre la Estructura de los Ocelos y vias Nerviosas Ocelares le Algunos Insectos, Trab. Lab. Invest. Biol. Univ. Madrid **16:**109. Refer to Harley Williams, *Don Quixote of the Microscope,* Jonathan Cape, London, 1954.

Calvin, M., 1962, Path of carbon in photosynthesis, *Science* **135:**879.

Calvin, M., 1969, *Chemical Evolution,* Oxford University Press, London.

Calvin, M., 1975, Chemical evolution, *Am. Sci.* **63:**169.

Calvin, M., 1983, Artificial photosynthesis quantum capture and energy storage, *Photochem. Photobiol.* **37:**349-360.

Chalazonitis, N., 1964, Light energy conversion in neuronal membranes, *Photochem. Photobiol.* **3:**539-559.

Chapman, D., 1973, Some recent studies of lipid, lipid-cholesterol and membrane systems, *Biol. Membr.* **2:**91.

Chapman, D., 1979, *Liquid Crystals,* F. D. Saeva, ed., Marcel Dekker, New York, pp. 305-334.

Chappell, R. L., and J. E. Dowling, 1972, Neural organization of the median ocellus of the dragonfly, *J. Gen. Physiol.* **60:**121.

Chase, R., 1979, Photic sensitivity of the rhinophore in *Aplysia, Can. J. Zool.* **57:**698-701.

Clarke, G. L., and E. J. Denton, 1962, Light and animal life, in *The Sea,* M. N. Hill, ed., Wiley (Interscience), New York, pp. 456-468.

Cohen, R., and M. Delbrück, 1959, Photoreactions in *Phycomyces,* growth and tropic responses to the stimulation of narrow test areas, *J. Gen. Physiol.* **42:**677.

Cohen, S. S., 1970, Are/were mitochondria and chloroplasts microorganisms?, *Am. Sci.* **58:**281.

Cohen, S. S., 1973, Mitochondria and chloroplast revisited, *Am. Sci.* **61:**437.

Cormier, M. J., ed., 1973, *Chemiluminescence and Bioluminescence,* Plenum, New York.

Crescitelli, F., 1972, The visual cells and pigments of the vertebrate, in *Handbook of Sensory Physiology,* H. J. A. Dartnall, ed., vol. 7, part 1, Springer-Verlag, New York, pp. 245-363.

Crescitelli, F., and H. J. A. Dartnall, 1953, Human visual purple, *Nature* **172:**195.

Croll, N. A., 1966, The phototactic response and spectral sensitivity of *Chromadorina viridis* (Nematoda: Chromadorida) with a note on the nature of paired pigment spots, *Nematol.* **12:**610-614.

Culkin, F., 1971, Constituents of Ocean Waters, in *Deep Oceans,* P. J. Herring and M. R. Clarke, eds., Praeger Publishers, New York, pp. 121-129.

Curry, G. M., and K. V. Thimann, 1961, Phototropism: The nature of the photoreceptor in higher and lower plants, in *Photobiol., Third Int. Congr. Proc., 1960,* pp. 127-134.

Dartnall, H. J. A., 1957, *The Visual Pigments,* Wiley, New York.

Darwin, C., 1859, *On the Origin of Species by Means of Natural Selection, or the Preservation of Favored Races in the Struggle for Life,* Murray, London (republished by Modern Library, New York, 1936).

Darwin, C., 1892, Letter to Joseph Hooker, 1871, in *The Autobiography of Charles Darwin and Selected Letters,* F. Darwin, ed., Appleton, New York (republished by Dover, New York, 1958), p. 220.

De Duve, C., 1984, *A Guided Tour of the Living Cell,* vols. 1 and 2, W. H. Freeman, New York.

De Duve, C., B. C. Pressman, R. Gianetto, R. Wattiaux, and F. Applemans, 1955, Tissue fractionation studies. 6. Intracellular distribution patterns of enzymes in rat-liver tissue, *Biochem. J.* **60:**604.

Delbrük, M., and W. Shropshire, Jr., 1960, Action and transmission spectra of *Phycomyces, Plant Physiol.* **35:**194.

Denton, E. J., J. B. Gilpen-Brown, and P. G. Wright, 1970, On the filter in the photophores of mesopelagic fish and on a fish emitting red light and especially sensitive to red light, *J. Physiol.* (London) **208:**72.

Denton, E. J., J. B. Gilpen-Brown, and P. G. Wright, 1972, The angular distribution of the light produced by some mesopelagic fish in relation to their camouflage, Roy. Soc. Proc., Ser. B **182:**145.

Denton, E. J., P. J. Herring, E. A. Widder, M. F. Latz, and J. F. Case, 1985, The roles of filters in the photophores of oceanic animals and their relation to vision in the oceanic environment, *R. Soc. London Proc.,* ser B, **225:**63-97.

Descartes, R., 1637, *Descartes: Discourse on Method, Optics, Geometry and Meterology,* P. J. Olscamp, transl., A. Leyde, De l'imprimerie de I, Maine (republished by Bobbs-Merrill, New York, 1965). Refer to Zrenner, 1985.

Dethier, V. G., 1962, *To Know a Fly,* Holden-Day, San Francisco, Calif.

Dickerson, R. E., 1971, Sequence and structure homologies in bacterial and mammalian type cytochromes, *J. Mol. Biol.* **57:**1.

Dilly, P. N., and J. J. Wolken, 1973, Studies on the receptors in *Ciona intestivalis.* IV. The ocellus of the adult, Micron **4:**11-29.

Dobell, C., 1958, *Antony van Leeuwenhoek and His 'Little Animals',* Russell & Russell, New York.

Dowling, J. E., and H. Ripps, 1970, Visual adaptation in the retina of the skate, *J. Gen. Physiol.* **56:**491.

Dubois, R., 1885, Note sur la physiologie des pyrophores, C. R. Soc. Biol., Ser. 8 **2:**559.

Duke-Elder, S., 1958, *System of Ophthamology: The Eye in Evolution,* vol. 1, Mosby, St. Louis, Mo.

Dumortier, B., 1972, Photoreception in the circadian rhythm of stridulatory activity in *Ephippiger, J. Comp. Physiol.* **77:**81-112.

Dupre, D. B., and E. T. Samulski, 1979, The fourth state of matter, in *Liquid Crystals,* F. D. Saeva, ed., Marcel Dekker, New York, pp. 203-247.

Duysens, L. N. M., 1964, Photosynthesis, *Prog. Biophys., Mol. Biol.* **14:**2-104.

Eakin, R. M., 1965, Evolution of photoreceptors, *Cold Spring Harbor Symp. Quant. Biol.* **30:**363.

Eakin, R. M., 1973, *The Third Eye,* University of California Press, Berkely, Calif.

Ebrey, T. S., 1967, Fast light-evoked potential from leaves, *Science* **155:**1556.

Eguchi, E., and T. H. Waterman, 1966, Fine structure patterns in crustacean rhabdoms, in *Functional Organization of the Compound Eye,* C. G. Bernhard, ed., Pergamon Press, Oxford, pp. 105-124.

Ehrenkranz, J. R. L., 1983, A gland for all seasons, *Nat. Hist.* **6:**19-23.

Einstein, A., 1905, On a heuristic viewpoint concerning the production and transformation of light, *Ann. Phys.* (Leipzig) **17:**132.

Eldred, W. D., and J. Nolte, 1978, Pineal photoreceptors: Evidence for a vertebrate visual pigment with two physiologically active states, *Vis. Res.* **18:**29-32.

Emerson, R., 1956, Effect of temperature on the long-wave limit of photosynthesis, *Science* **123:**637.

Emerson, R., and C. M. Lewis, 1943, The dependence of the quantum yield of *Chlorella* photosynthesis on wavelength of light, *Am. J. Bot.* **30:**165.

Englemann, T. W., 1882, Uber Licht- und Farbenperception neiderster Organismen, Pflügers Arch. Gesamte Physiol., Menschen Tiere **29:**387.

Etkin, W., 1973, Structure of the DNA molecule, *Biosci.* **23:**653.

Exner, S., 1891, *Die Physiologie der Facettierten Augen von Krebsen und Insekten,* Deuticke, Vienna.

Favre, A., and G. Thomas, 1981, Transfer RNA: From photophysics to photobiology, *Ann. Rev. Biophys. Bioeng.* **10:**175-195.

Fergason, J. L., and G. H. Brown, 1968, Liquid crystals and living systems, *J. Am. Oil Chem. Soc.* **45:**120.

Fernel, Jean, 1542, Physiology IV. C. I. Dialog ii C. 7, quote from *The Endeavor of Jean Fernal,* Charles Sherrington, 1946, Cambridge University Press, New York, p. 39.

Fingerman, M., 1960, Tidal rhythmicity in marine organisms, *Cold Spring Harbor Symp. Quant. Biol.* **25:**481.

Fischer, H., and A. Stern, 1940, *Die Chemie des Pyrrols,* vol. 2, halfte 2, Akad. Verlagsges, Leipzig.

Fitch, W. M., and E. Margoliash, 1967, Construction of phylogenetic trees, *Science* **155:**279.

Foster, K. W., and R. D. Smyth, 1980, Light antennas in phototactic algae, *Microbiol. Rev.* **44:**572-630.

Foster, K. W., J. Saranak, N. Patal, G. Zarilli, M. Okabe, T. Kline, and K. Nakanishi, 1984, A rhodopsin is the functional photoreceptor for phototaxis in the unicellular eukaryote *Chlamydomonas, Nature* **311:**756.

Fox, D. L., 1953, *Animal Biochromes and Structural Colors,* Cambridge University Press, London; 2nd ed., 1976 Univ. of California Press, Los Angeles, Calif.

Fox, D. L., 1960, Pigments of plant origin in animal phyla, in *Comparative Biochemistry of Photoreactive Systems,* M. B. Allen, ed., Academic Press, New York, pp. 11-31.

Fox, S. W., 1965*a,* A theory of macromolecular and cellular origins, *Nature* **205:**328.

Fox, S. W., 1965*b, The Origins of Prebiological Systems,* Academic Press, New York.

Fox, S. W., 1980, Metabolic microspheres: Origin and evolution, *Nat. wiss.* **67:**378-383.

Fox, S. W., K. Harada, K. R. Woods, and C. R. Windsor, 1963, Amino acids composition of proteinoids, *Arch. Biochem. Biophys.* **102:**439.

Franz, V., 1931, Die Akkomodation des Selachierauges und seine Abblendungsapparate nebst Befunden an der Retina, Zool. Jahrb., *Abt. Allg. Zool. Physiol. Tiere* **49:**323.

Franz, V., 1934, Vergleichende Anatomie des Wirbeltierauges, in *Handbook der vergleichenden Anatomie der Wirbeltiere,* L. Bolk et al., eds., vol. 1, part 2, Urban & Schwarzenberg, Berlin, pp. 1009-1023.

von Frisch, K., 1949, Die Polarization des Himmelslichtes als Faktor der Orientieren bei den Tanzen der Bienen, *Experientia* **5:**397.

von Frisch, K., 1950, *Bees: Their Vision, Chemical Senses and Language,* Cornell University Press, Ithaca, N.Y., revised edition 1971.

von Frisch, K., 1967, *The Dance Language and Orientation of Bees,* Belknap Press, Cambridge, Mass.

Gaffron, H., 1965, The role of light in evolution: The transition from a one quantum to a two quantum mechanism, in *The Origins of Prebiological Systems,* S. W. Fox, ed., Academic Press, New York, pp. 437-460.

Garten, S., 1907, Die Veranderungen der Netzhaut durch Licht, in *Graefe Saemisch Handbuch der gesamten Augenheilkunde,* Sulf. 2, Bd. 3, Kap 12, Leipzig, pp. 1-130.

Gatenby, J., and L. Ellis, 1951, The vertebrate neuron, the pituitary, and sudan black, *Cellule* **54:**149.

Geethabali, X., and K. P. Rao, 1973, A metasomatic neural photoreceptor in the scorpion, *J. Exp. Biol.* **58:**189-196.

Gilbert, P. W., 1963, The visual apparatus of sharks, in *Sharks and Survival,* P. W. Gilbert, ed., Heath, Lexington, Mass., pp. 283-326.

Glover, J., T. W. Goodwin, and R. A. Morton, 1948, Conversion in vivo of Vitamin A aldehyde (retinene$_1$) to Vitamin A$_1$, *Biochem. J.* **43:**109.

Goldsmith, T. H., 1958a, The visual system of the honeybee, Nat. Acad. Sci. U.S.A. Proc. **44:**123.

Goldsmith, T. H., 1958b, On the visual system of the bee *(Apis mellifera),* N.Y. Acad. Sci. Ann. **74:**223.

Goldsmith, T. H., 1972, The natural history of invertebrate visual pigments, in *Handbook of Sensory Physiology,* H. J. A. Dartnall, ed., vol. 7, part 1, Springer-Verlag, Berlin, pp. 685-719.

Gotow, T., 1975, Morphology and function of photoexcitable neurons in the central ganglia of *Onchidium verruculatum, J. Comp. Physiol.* **99:**139-157.

Granick, S., 1948, Magnesium protoporphyrin as a precursor of chlorophyll in *Chorella, J. Biol. Chem.* **175:**333.

Granick, S., 1950, Magnesium vinyl pheoporphyrin a$_5$, another intermediate in the biological synthesis of chlorophyll, *J. Biol. Chem.* **183:**713.

Granick, S., 1958, Porphyrin biosynthesis in erythrocytes. I. Formation of delta-amino-levulinic acid in erythrocytes, *J. Biol. Chem.* **232:**1101.

Greef, R., 1877, *Monograph Alciopidae,* Nova Acta Kais, Leopold *39,* No. 2.

Gregory, R. L., 1966, *Eye and Brain,* McGraw-Hill, New York.

Gregory, R. L., 1967, Origin of eyes and brains, *Nature* **213:**369.

Gregory, R. L., N. Moray, and H. E. Ross, 1964, The curious eye of Copilia, *Nature* **201:**1166.

Grenacher, H., 1879, *Untersuchungen über das Sehorgan der Arthropoden, insbesondere der Spinner, Insekten and Crustaceen,* Vanderhoeck and Ruprecht, Gottingen, Germany, p. 195.

Grenacher, H., 1886, Abhandlungen zur vergleichen Anatomie des Auges I. Die Retina der Cephalopoden, *Abh. Naturforsch. Ges. Halle* **16:**207.

Gruber, S. H., D. I. Hamasaki, and C. D. B. Bridges, 1963, Cones in the retina of the lemon shark *(Negaprion brevirotris), Vis. Res.* **3:**397.

Gurwitsch, A. A., 1968, *Problems of Mitogenetic Radiation as an Aspect of Molecular Biology,* Meditsina, Leningrad, USSR.

Haldane, J. B. S., 1928, *Possible World,* Harper, New York.

Haldane, J. B. S., 1954, *The Biochemistry of Genetics,* Macmillan, New York.

Haldane, J. B. S., 1966, *The Causes of Evolution,* Cornell University Press, Ithaca, N.Y.

Hamasaki, D. I., and P. Streck, 1971, Properties of the epiphysis cerebri of the small spotted dogfish shark *Scyliorhinus caniculus* L., *Vis. Res.* **11:**189-198.

Hanoaka, T., and K. Fujimoto, 1957, Absorption spectrum of a single cone in the carp retina, *Jpn. J. Physiol.* **7:**276.

Hara, T., and R. Hara, 1980, Retinochrome and rhodopsin in the extraocular photoreceptor of the squid *Todarodes, J. Gen. Physiol.* **75:**1-19.

Harth, M. S., and M. B. Heaton, 1973, Nonvisual photoresponsiveness in newly hatched pigeons *(Columba livia), Science* **180:**753-755.

Hartmann, K. M., 1966, A general hypothesis to interpret high energy phenomena of photomorphogenesis on the basis of phytochrome, *Photochem. Photobiol.* **5:**349.

Hartwig, H. G., and C. Baumann, 1974, Evidence for photosensitive pigments in the pineal complex of the frog, *Vis. Res.* **14:**597-598.

Harvey, E. N., 1952, *Bioluminescence,* Academic Press, New York.

Harvey, E. N., 1960, Bioluminescence, *Comp. Biochem.* **2:**545-591.

Hastings, J. W., 1968, Bioluminescence, in *Annual Review of Biochemistry,* P. D. Boyer, ed., **37,** pp. 597-630.

Hawkins, E. G. E., and R. F. Hunter, 1944, Vitamin A aldehyde, *J. Chem. Soc.* (London), *411.*

Heller, J., 1969, Comparative study of a membrane protein characterization of bovine, rat and frog visual pigment, Biochem. **8:**675.

von Helmholtz, H., 1852, Über die Theorie der zusammengesetzten Farben, *Ann. Phys.* (Leipzig) (2) **87:**45.

von Helmholtz, H., 1867, *Handbuch der Physiologischen Optik,* Voss, Leipzig.

Hendricks, S. B., 1968, How light interacts with living matter, *Sci. Am.* **219:**174.

Hendricks, S. B., and H. W. Seigleman, 1967, Phytochrome and Photoperiodism in plants, *Comp. Biochem.* **27:**211-235.

Hering, E., 1885, Ueber Individuelle Verachiedenbeiten des Farbensinnes, Lotos, Prague, and in *Outlines of a Theory of the Light Sense,* L. M. Hurvich and D. Jameson, transl.), Harvard University Press, Cambridge, Mass., 1965.

Hermans, C. O., and R. M. Eakin, 1974, Fine structure of eyes of an alciopid polychaete, *Vanadis tangensis* (Annelida), *Z. Morphol., Tierre* **79:**245-267.

Hesse, R., 1899, Untersuchungen über die Organe der Lichtenpfindung bie niederen Thieren. V. Die Augen den polychaten Anneliden, *A. Wiss. Zool.* **65:**446-516.

Hill, R., and F. Bendall, 1960, Function of the two cytochrome components in chloroplasts: A working hypothesis, *Nature* **186:**136.

Hisano, N., H. Tateda, and M. Kubara, 1972*a,* Photosensitive neurons in the marine plumonate mollusk *Onchidium verruculatum, J. Exp. Biol.* **57:**651-660.

Hisano, N., H. Tateda, and M. Kubara, 1972*b,* An electrophysiological study of the photo-excitable neurons of *Onchidium verruculatum* in situ, *J. Exp. Biol.* **57:**661-671.

Hisano, N., D. P. Cardinalli, J. M. Rosner, C. A. Nagle, and J. H. Tremezzani, 1972, Pineal role in the duck extraretinal photoreception, Endocrinol. **91:**1318-1322.

Hooke, R., 1665, *Micrographia,* reprinted in *Early Science in Oxford,* R. T. Gunther, 1938, Oxford University Press, Oxford.

Hooke, R., 1705, *Posthumous Works of Robert Hooke,* Richard Waller, London.

Horridge, G.A., 1975, *The Compound Eye and Vision of Insects,* Oxford University Press, New York.

Hubbard, R., P. K. Brown, and D. Bownds, 1971, Methodology of Vitamin A and visual pigments, in *Methods in Enzymology,* D. B. McCormick and L. D. Wright, eds., vol. 18C, Academic Press, New York, pp. 615-653.

Hunter, R. F., and N. E. Williams, 1945, Chemical conversion of beta-carotene into Vitamin A, *J. Chem. Soc.* (London), part 2, 554-555.

Hydén, H., 1967, Biochemical and molecular aspects of learning and memory, Am. Phil. Soc. Proc. **111:**326.

Hydén, H., and P. W. Lange, 1968, Protein synthesis in the hippocampal pyramidal cells of rats during a behavioral test, *Science* **159:**1370.

Ingen-Housz, J., 1779, *Experiments upon Vegetables, Discovering their Great Power of Purifying the Common Air in Sunshine and Injuring It in the Shade and at Night,* Elmsly & Payne, London.

Ingen-Housz, J., 1798, *Ernabrung der Pflanzen and Fruchtbarkeit des Bodens,* Leipzig.

Johnson, C. H., and J. W. Hastings, 1986, The elusive mechanism of the circadian clock, *Am. Sci.* **74:**29-36.

Johnson, F. H., 1967, Bioluminescence, *Comp. Biochem.* **27:**79-136.

Journeaux, R., and R. Viovy, 1978, Orientation of chlorophylls in liquid crystals, *Photochem. Photobiol.* **28:**243.

Judson, H. F., 1979, *Eighth Day of Creation,* Simon and Schuster, New York.

Karnaukhov, V. N., 1973, The nature and function of yellow aging pigment, lipofuscin, *Exp. Cell Res.* **80:**479.

Ke, B., and L. Vernon, 1971, Living systems in photochromism, in *Photochromism,* G. H. Brown, ed., Wiley (Interscience), New York, p. 687.

Keeble, F., 1910, *Plant-Animal: A Study in Symbiosis,* Cambridge University Press, New York.

Kelly, D., 1965, Ultrastructure and development of amphibian pineal organs, Prog. Brain Res. **10:**270.

Kelly, D., 1971, Developmental aspects of amphibian pineal systems, *Pineal Gland, Ciba Found. Symp.,* 1970, pp. 53-77.

Kelner, A., 1949, Effect of visible light on the recovery of *Streptomyces griseus* conidia from ultraviolet irradiation injury, Nat. Acad. Sci. U.S.A. Proc. **35:**73.

Keyes, R. W., 1985, Optical—in the light of computer technology, *Optica Acta* **32:**525-535.

Kirlian, Semyon D., and V. Kirlian, 1961, Photography and visual observations by means of high frequency currents, *J. Sci. Appl. Photog.* **6:**397-403. Refer also to Tiller, W. A., ed., 1973, *Preceedings of the First Hemisphere Conference on Kirlian Photography,* Gordon and Breach, N.Y., and to Boyers, D. G., and W. A. Tiller, 1973, Corona discharge photography, *J. Appl. Phys.* **44:**3102.

Kraughs, J. M., L. A. Sordhal, and A. M. Brown, 1977, Isolation of pigment granules involved in extra-retinal photoreception in *Aplysia californica* neurons, *Biochim. Biophys. Acta* **471:**25-31.

Krause, W., 1863, Uber die Endigung der Muskelnerven, *Z. Ration. Med.* **20:**1.

Krieger, D., 1983, Brain peptides; what, where, and why? *Science* **222:**975-985.

Kühn, H., 1968, On possible ways of assembling simple organized systems of molecules, in *Structural Chemistry and Molecular Biology,* A. Rich, and N. Davidson, eds., Freeman, San Francisco, Calif., pp. 566-572.

Kühne, W., 1878, *The Photochemistry of the Retinal and on Visual Purple*, M. Foster, ed., Macmillan, New York.

Kuhnert, L., 1986, A new optical photochemical memory device in a light-sensitive chemical medium, *Nature* **319**:393-394.

Land, M. F., 1980, Optics and vision in invertebrate, in *Handbook of Sensory Physiology*, H. Autrum, ed., vol.7/6B, Springer-Verleg, Berlin, pp. 472-592.

Langer, H., 1967, Über die Pigmentgranula im Facettenaugen von *Callifora erythrocephala*, *A. Vgl. Physiol.* **55**:354.

Langer, H., and B. Thorell, 1966, Microspectrophotometry of single rhabdomeres in the insect eye, *Exp. Cell Res.* **44**:673.

Lavoisier, A. L., 1774, *Opuscules Physique et Chimiques*, Durand, Paris.

Leduc, S., 1911, *Mechanisms of Life*, London.

Lee, A. G., 1975, Interactions within biological membranes, Endeav. **34**:67-71.

van Leeuwenhoek, A. (1674). More observation from Mr. Leeuwenhoek, *R. Soc. London Phil. Trans.* **9**:178. (See Dobell, 1958).

Lehmann, O., 1904, *Füssige Kristalle, sowie Plästizitat von Kristallen im Allgemeinin, Moleculare Umlagerungen und Aggregatzumstandsänderugen*, Englemann, Leipzig.

Lehman, O., 1922, *Handbuch der biologischen Arbeitsmethoden*, Physik-Chem. Methoden, Untersuchung des Verhaltens gelöster Stoffe, E. Arberhalden, ed., AB. III, Teil A2, Urban and Schwarzenberg, Munich, pp. 123-352.

Lehninger, A. L., 1965, *Bioenergetics*, Benjamin, N.Y., pp. 222-224.

Lentz, T. L., 1968, *Primitive Nervous Systems*, Yale University Press, New Haven, Conn., pp. 112-113.

Lerner, A. B., J. D. Case, Y. Takahashi, T. H. Lee, and W. Mori, 1958, Isolation of melatonin, the pineal gland factor that lightens melanocytes, *J. Am. Chem. Soc.* **80**:2587.

Lewin, R. A., 1962, *The Physiology and Biochemistry of Algae*, Academic Press, New York.

Lickey, M. E., and S. Zack, 1973, Extraocular photoreceptors can entrain the circadian rhythm in the abdominal ganglion of *Aplysia, J. Comp. Physiol.* **86**:361-366.

Liebman, P., and G. Entine, 1964, Sensitive low-light level of microspectrophotometric detection of photosensitive pigment of retinal cones, *J. Opt. Soc. Am.* **54**:1451.

Lightner, D.A., and A. F. McDonagh, 1984, Molecular mechanisms of phototherapy for neonatal jaundice, *Acc. Chem. Res.* **17**:417-424.

Lipetz, L. E., 1984, Pigment types, densities and concentrations in cone oil droplets of *Emydoidea Blandingii, Vision Res.* **24**:605.

Loew, E. R., and J. N. Lythgoe, 1985, The ecology of color vision, *Endeavor* **14**:170-174.

Lubbock, Sir J., 1882, *Ants, Bees and Wasps*, Appleton, New York.

Lumière, A., and L. Lumière, 1894, Note sur la photographie des couleurs. Académie des Science de Lyon, in *Principaux Travaux Scientific Auguste Lumière*, Leon Sezanne, Lyon, 1928.

Lynch, G., and M. Baundry, 1984, The biochemistry of memory: A new and specific hypothesis, *Science* **224**:1057-1063.

MacNichol, E. F., 1964, Three pigment color vision, *Sci. Am.* **211**:48.

Macullum, A. B., 1926, The paleochemistry of the body fluids and tissues, *Physiol. Rev.* **6**:316.

Manning, J. E., and O. C. Richards, 1972, Synthesis and turnover *Euglena gracilis* nuclear and chloroplast deoxyribonucleic acid, Biochem. **11**:2036.

Marak, G. E., and J. J. Wolken, 1965, An action spectrum of the fire ant: *Solenopsis saevissima, Nature* **205:**1328.

Margulis, L., 1970, *Origin of Eucaryotic Cells,* Yale University Press, New Haven, Conn.

Margulis, L., 1982, *Early Life,* Science Books International, Boston, Mass.

Marks, P. S., 1976, Nervous control of light responses in the sea anemone, *Calamactis praelongus, J. Exp. Biol.* **65:**85-96.

Marks, W. B., 1963, Difference spectra of the visual pigments in single goldfish cones, Thesis, Johns Hopkins University, Baltimore, Md.

Marks, W. B., W. H. Dobele, and J. B. MacNichol, 1964, Visual pigments of single primate cones, *Science* **143:**1181.

Mast, S. O., 1911, *Light and the Behavior of Organisms,* Wiley, New York.

Mauro, A., 1977, Extraocular photoreceptors in cephalopods, *Symp. Zool. Soc. Lond.* **38:**287-308.

Mauro, A., and F. Baumann, 1968, Nervous control of light responses of photoreceptors in the epistellar body of *Eledone moschata, Nature* **220:**1332-1334.

Mauro, A., and A. Sten-Knudsen, 1972, Light-evoked impulses from extraocular photoreceptors in the squid *Todarodes, Nature* **237:**342-343.

Maxwell, J. C., 1853, in *Scientific Papers of J. Clerk Maxwell,* Vol. I, W. D. Niven, ed., Dover Books, New York, 1952.

Maxwell, J. C., 1861, The theory of compound colours and the relations of the colours of the spectrum, Roy. Soc. London Phil. Trans. **150:**57.

Maxwell, J. C., 1890, The theory of compound Colours and the relation of the colours of the spectrum, in *The Scientific Papers of J. Clerk Maxwell,* Vol. I, Cambridge University Press, London, pp. 126-154, reprinted and edited by W. D. Niven, Dover Press, New York, 1952.

Mayer, R., 1845, *Die Organische Bewegung in ihrem Zusammenhang mit dem Stoffwechsel,* Heilbronn.

McCapra, F., 1973, The chemistry of bioluminescence, Endeav. **32:**139.

McConnell, J. V., 1962, Memory transfer through cannibalism in planarians, *J. Neuropsychiatr.* **3:**Suppl. 1, 42.

McConnell, J. V., 1966, Comparative physiology: Learning in invertebrates, *Ann. Rev. Physiol.* **28:**107.

McConnell, J. V., and J. M. Shelby, 1970, Memory transfer in invertebrates, in *Molecular Mechanisms in Memory and Learning,* G. Ungar, ed., Plenum, New York, pp. 71-101.

McConnell, J. V., A. L. Jacobson, and D. P. Kimble, 1959, The effects of regeneration upon retention of a conditional response in the planarian, *J. Comp. Physiol. Psychol.* **52:**1.

McCrone, E. J., and P. G. Sokolove, 1979, Brain-gonad axis and photoperiodically stimulated sexual maturation in the slug, *Limax maximus, J. Comp. Physiol.* **133:**117-123.

McElroy, W. D., H. H. Seliger, and E. H. White, 1969, Mechanism of bioluminescence, chemiluminescence and enzyme function in the oxidation of firefly luciferin, *Photochem. Photobiol.* **10:**153.

McMillan, J. P., J. A. Elliot, and M. Menaker, 1975a, On the role of eyes and brain photoreceptors in the sparrow: A Schoff's rule, *J. Comp. Physiol.* **102:**257-262.

McMillan, J. P., J. A. Elliot, and M. Menaker, 1975*b,* On the role of eyes and brain photoreceptors in the sparrow: Arrhythmicity in constant light, *J. Comp. Physiol.* **102:**263-268.

Meissner, G., and M. Delbrück, 1968, Carotenes and retinal in *Phycomyces* mutants, *Plant Physiol.* **43:**1279.

Menaker, M., ed., 1976, Extra-retinal photoreception, *Photochem. Photobiol.* **23**(4):213-306.

Menaker, M., 1977, Extra-retinal photoreceptors, in *The Science of Photobiology,* K. C. Smith, ed., Plenum, New York, pp. 227-240.

Menger, E. L., ed., 1975, Chemistry of vision, in *Accounts of Chemical Research 8,* no. 3, American Chemical Society, Washington, D.C., pp. 81-112.

Meyer, J. R., 1977, Head capsule transmission of long-wavelength light in the *Curculionidae, Science* **196:**524-525.

Miller, S. L., 1957, The mechanism of synthesis of amino acids by electric discharges, *Biochim. Biophys. Acta* **23:**480.

Miller, W. H., and G. D. Bernard, 1968, Butterfly glow, *J. Ultrastruct. Res.* **24:**286.

Millott, N., 1968, The dermal light sense, in *Invertebrate Receptors,* J. D. Carthy and G. E. Newell, eds., Academic Press, New York, pp. 1-36.

Millott, N., 1978, *Extra-ocular Photosensitivity,* Meadowfield Press, Burham, England.

Mitchell, P., 1966, Chemiosmotic coupling in oxidative and photosynthetic phosphorylation, Cambridge Phil. Soc. Biol. Rev. **41:**445.

Mogus, M. A., and J. J. Wolken, 1974, *Phycomyces:* Electrical response to light stimuli, *Plant Physiol.* **53:**512.

Molyneux, W., 1709, *Dioptrica Nova,* 2nd ed., printed for B. Tooke, London.

Moray, N., 1972, Visual mechanisms in the copepod *Copilia,* Percept. **1:**193.

Morison, W. L., 1984, Photoimmunology, *Photochem. Photobiol.* **40:**781-787.

Morton, R. A., 1944, Chemical aspects of the visual process, *Nature* **153:**69.

Morton, R. A., and T. W. Goodwin, 1944, Preparation of retinene in vitro, *Nature* **153:**405.

Munk, D., 1966, Ocular anatomy of some deep-sea teleosts, Dana Reports, *Photobiol.* no. 70, Host and Son, Copenhagen, pp. 60-62.

Munk, D., 1984, Non-spherical lenses in the eyes of deep-sea teleosts, *Arch. Fisch. wiss.* **34:**145-153.

Needham, J., 1950, *Biochemistry and Morphogenesis,* Cambridge University Press, New York

Needham, J., 1956, *Science and Civilization in China,* vol. 2, Cambridge University Press, London, p. 298.

Newton, Sir I., 1666, *Optics: A Tretise of the Reflections, Refractions, Inflections and Colors of Light,* reprinted McGraw-Hill, New York, 1931.

Ninneman, H., 1980, Blue light photoreceptors, *Biosci.* **30:**166-170.

van Niel, C. B., 1941, The bacterial photosynthesis and their importance for the general problem of photosynthesis, *Adv. Enzymol.* **1:**263.

van Niel, C. B., 1943, Biochemical problems of the chemo-autotrophic bacteria, *Physiol. Rev.* **23:**338.

van Niel, C. B., 1949, The comparative biochemistry of photosynthesis, *Am. Scientists* **37:**371.

O'Benar, J. D., and Y. Matsumoto, 1976, Light induced neural activity and muscle contraction in the marine worm, *Golfingia gouldii, Comp. Biochem. Physiol.* **55A:**77-81.

Oesterhelt, D., and W. Stoeckenius, 1971, Rhodopsin-like protein from the purple membrane of *Halobacterium halobium, Nature* **233:**149-152.

Ogawa, T., Y. Inoue, M. Kitajima, and K. Shibata, 1973, Action spectra for biosynthesis of chlorophyll a and b and beta-carotene, *Photochem. Photobiol.* **18:**229.

Ohtsuka, T., 1985, Relation of spectral types of oil droplets in cones of turtle retina, *Science* **229:**874-877.

Okada, Y. K., 1966, Observations on rod-like contents in the photogenic tissues of *Watasenia scintillons* through the electron microscope, in *Bioluminescence in Progress,* F. H. Johnson and Y. Haneda, eds., Princeton University Press, Princeton, N.J., pp. 611-625.

Olby, R., 1974, *The Path to the Double Helix,* University of Washington Press, Seattle, Wash.

Ootaki, T., and J. J. Wolken, 1973, Octahedral crystals in *Phycomyces* II, *J. Cell Biol.* **57:**278.

Oparin, A. I., 1938, *The Origin of Life,* S. Morgulis, transl., Macmillan, New York, 2nd ed., 1953, Dover, New York.

Oparin, A. I., 1968, *Genesis and Evolutionary Development of Life,* Academic Press, New York.

Oro, J., 1980, Prebiological synthesis of organic molecules and origin of life, in *The Origins of Life and Evolution,* H. O. Halvorson and K. E. Van Holde, eds., vol. I, Alan R. Liss, New York, pp. 47-63.

Packard, A., and G. Sanders, 1969, What the octopus shows the world, *Endeav.* **28:**92.

Palmer, J. D., 1974, *Biological Clocks in Marine Organisms,* Wiley, New York.

Park, R. B., and J. Biggins, 1964, Quantasome: size and composition, *Science* **144:**1009.

Pasteur, L., 1860, De l'origine des ferments. Nouvelle expériences relative aux generations dites spontanees, Adad. Sci. C. R. **50:**849.

Pasteur, L., 1878, *Collected Works of Pasteur (Oeuvres de Pasteur)* by Vallery-Radot (1922-1939), vols. 1 and 2, Masson, Paris.

Pauling, L., 1960, *The Nature of the Chemical Bond,* 3rd. ed., Cornell University Press, Ithaca, N.Y.

Pelletier, J., and J. B. Caventou, 1818, Sur la matiere verte des feuilles, *Ann. Chim. Phys.* **9:**194.

Planck, M., 1922, *The Origin and Development of the Quantum Theory,* H. T. Clarke and L. Silberstein, transl., Oxford University Press, Oxford.

Polyak, S., 1957, *The Vertebrate Visual System,* H. Kluver, ed., University of Chicago Press, Chicago.

Ponnamperuna, C., 1972, *The Origins of Life,* Dutton, New York.

Priestly, J., 1772, Observations on different kinds of air, Roy. Soc. London Phil. Trans. **62:**147.

Purkinje, J., 1825, *Beobachtungen und Versuche zur Physiologie der Sinne,* vol. 2, Reimer, Berlin.

Quickenden, T. I., and S. S. Que Hee, 1981, Speculations, *Sci. Technol.* **4:**453-464.

Quickenden, T. I., M. J. Comarond, and R. N. Tilbury, 1985, Ultraweak biolumines-cence of stationary phase *Saccharomyces cerevisae* and *Schizosaccharomyces pombe, Photochem. Photobiol.* **41:**611-615.

Rabinowitch, E., 1945, *Photosynthesis and Related Processes,* vol. 1, Wiley (Interscience), New York.

Rabinowitch, E., 1951, *Photosynthesis and Related Processes,* vol. 2, part 1, Wiley (Interscience), New York.

Rabinowitch, E., 1956, *Photosynthesis and Related Processes,* vol. 2, part 2, Wiley (Interscience), New York.

Rayport, S., and G. Wald, 1978, Frog skin photoreceptors, *Am. Soc. Photobiol. Abstr.* **6:**94-95.

Reich, W., 1948, *The Discovery of the Orgone,* Orgone Institute Press, New York.

Reinitzer, F. O., 1888, Beitrage zur Kenntis der Cholesterins, Monatsh. *Chem.* **9:**421.

Ribi, W. A., 1981, The phenomena of eye glow, *Endeavor* **5:**2-8.

Rinne, F., 1933, Investigations and considerations concerning paracrystallinity, *Faraday Soc. Trans.* **29:**1016.

Robinson, C., 1956, Liquid-crystalline structures in solution of a polypeptide, *Faraday Soc. Trans.* **52:**571.

Robinson, C., 1958, Liquid-crystalline structures in solution of a polypeptide, Part II, *Faraday Soc. Discussions* pp. 29-42.

Robinson, C., 1966, The cholesteric phase in polypeptides and biological structures, *Mol. Cryst.* **1:**467.

Rölich, P., and I. Törö, 1965, Fine structure of the compound eye of *Daphnia* in normal, dark- and strongly light-adapted state, in *The Eye Structure II Symposium,* J. W. Rohen, ed., F. K. Schattauer-Verlag, Stuttgart, pp. 175-186.

Rosenberg, A., 1967, Galactosyl Diglycerides: Their possible function in *Euglena* chloroplasts, *Science* **157:**1191.

Rüdiger, W., 1969, Über die struktur des Phytocrom-chromophors und seine Protein-Bindung, *Justus Liebigs. Ann. Chem.* **723:**208.

von Sachs, J., 1864, Wirkungen Farbigen Lichts and Pflanzen, *Botan. Z.* **22:**353.

Sager, R., 1972, *Cytoplasmic Gene and Organelles,* Academic Press, New York.

de Saussure, N., 1804, *Recherches Chimiques sur la Vegatation,* Lyon.

Scharrer, E., 1964, A specialized trophospongium in large neurons of *Leptodora* (Crustacea), *Z. Zellforsch. Mikrosk. Anat.* **61:**803.

Schleiden, M. J., 1838, *Beitrage zur Botanik,* Gesammelte Auffassa, Englemann, Leipzig.

Schopf, S. W., ed., 1983, *The Earth's Early Atmosphere: Its Origin and Evolution,* Princeton University Press, Princeton, N.J.

Schultze, M. J., 1866, Zur Anatomie und Physiologie der Retina, *Arch. Mikrosk. Anat.* **2:**175.

Schwann, T., 1839, *Mikroskopische Unteruchungen über die Ueber-stimmung in der Struktur und dem Wachsthum der Tiere and Pflanzen,* Sander, Berlin.

Seliger, H. H., and W. D. McElroy, 1965, *Light: Physical and Biological Action,* Academic Press, New York.

Senebier, J., 1782, *Memoires Physico-chimiques sur L'influence de la Lumiere Solaire pour Modifier les Etros de Trois Regnes, surtout ceux de Regne Vegetal,* 3 vols., Chirol, Geneve.

Serebrovskaya, I., 1971, *Chemical Evolution and the Origin of Life,* R. Buvet and C. Ponnamperuma, eds., American Elsevier, New York, pp. 297-306.

Shemin, D., 1948, The biosynthesis of porphyrins, *Cold Spring Harbor Symp. Quant. Biol.* **13:**185-192.

Shemin, D., 1955, The succinate-glycine cycle: The role of delta-amino-levulinic acid in porphyrin synthesis, *Porphyrin Biosyn. Metab., Ciba Found. Symp., 1955,* pp. 4-28.

Shemin, D., 1956, The biosynthesis of porphyrins, *Harvey Lect.* **50:**258-284.

Shimomura, O., F. H. Johnson, and Y. Saiga, 1963, Partial purification and properties of the *Osontosyllis* luminescence system, *J. Cell Comp. Physiol.* **61:**275-292.

Sibaoka, T., 1962, Excitable cells in Mimosa, *Science* **137:**226.

Slawinska, D., and J. Slawinski, 1983, Biological chemiluminescence, *Photochem. Photobiol.* **37:**709-715.

Smith, D. E., 1963, The organization and innervation of the luminescent organ in a firefly, *Photuris pennsylvanica* (Coleoptera), *J. Cell Biol.* **16:**323.

Sohol, R. S., 1981, *Age Pigments,* Elsevier, North Holland, Amsterdam, pp. 303-316.

Stanier, R. Y., 1959, Formation and function of the photosynthetic pigment in purple bacteria, *Brookhaven Symp. Biol.* **11:**43-53.

Steven, D. M., 1963, The dermal light sense, Cambridge Phil. Soc. Biol. Rev. **38:**204.

Stevens, J. K., and K. E. Parsons, 1980, A fish with double vision, *Nat. Hist.* **89:**62-67.

Strehler, B. L., 1962, *Time, Cells and Aging,* Academic Press, New York.

Strother, G. K., 1963, Absorption spectra of retinal oil globules in turkey, turtle, and pigeon, *Exp. Cell Res.* **29:**249.

Strother, G. K., and A. J. Casella, 1972, microspectrophotometry of arthropod visual screening pigments, *J. Gen. Physiol.* **59:**616.

Strother, G. K., and J. J. Wolken, 1960, Microspectrophotometry, I. Absorption spectra of colored oil globules in the chicken retina, *Exp. Cell Res.* **21:**504.

Stuermer, W., 1970, Soft parts of cephalopods and trilobites: Some surprising results of X-ray examinations of Devonian slates, *Science* **170:**1300.

Sweeney, B. M., D. C. Fork, and K. Satoh, 1983, Stimulation of bioluminescence in dinoflagellates by red light, *Photochem. Photobiol.* **37:**457-465.

Tamarkin, L., C. J. Baird, O. F. X. Almeida, 1985, Melatonin: A coordinating signal for mammalian reproduction, *Science* **227:**714-727.

Tomita, T., 1970, Electrical activity of vertebrate photoreceptors, *Q. Rev. Biophys.* **3:**179.

Towe, K. M., 1973, Trilobite eyes calcified lenses in vivo, *Science* **179:**1007.

Trainor, G. L., 1979, Studies on the *Odontosyllis* bioluminescence systems, Thesis, Harvard University, Cambridge, Mass.

Truman, J. W., 1976, Extraretinal photoreception in insects, *Photochem. Photobiol.* **23:**215-225.

Ungar, G., 1970, Role of proteins and peptides in learning and memory, in *Molecular Mechanisms in Memory and Learning,* G. Ungar, ed., Plenum, New York, pp. 149-175.

Urey, H., 1952, *The Planets,* Yale University Press, New Haven, Conn.

Vaissière, R., 1961*a,* Morphologie et Histologie Comparées des Yeux des Crustacés Copépodes, Thesis, Centre National de la Recherche Scientifique, Paris.

Vaissière, R., 1961*b,* Morphologie et Histologie Comparées des Yeux des Crustacés Copépodes, *Arch. Zool. Exp. Gen.* **100:**126.

Van Brunt, E. E., M. D. Sheperd, J. R. Wall, W. F. Ganong, and M. T. Clegg, 1964, Penetration of light into the brain of mammals, *N. Y. Acad. Sci. Ann.* **117:**204-216.

Van Veen, T., H. G. Hartwig, and K. Muller, 1976, Light-dependent motor activity and photonegative behavior in the eel, *J. Comp. Physiol.* **11A:**209-219.

Vigroux, E., 1969, Coefficiente d'absorption de l'ozone dans la bande de Hartley, *Geophys. Ann.* **25:**169.

Virchow, R., 1858, *Die Cellularpathologie in Ihrer Begrundung auf Physiologische und Pathologische Gewebelehre,* Hirschwald, Berlin.

Vogt, K., 1977, Ray path and reflection mechanisms in crayfish eyes, *Z. Nat.forsch.* **32C:**466-468.

Wald, G., 1940, The distribution of vitamin A_1 and A_2, *J. Gen. Physiol.* **22:**301.

Wald, G., 1948, Galloxanthin, a carotenoid from the chicken retina, *J. Gen. Physiol.* **31:**377.

Wald, G., 1953, The biochemistry of vision, *Annu. Rev. Biochem.* **22:**497.

Wald, G., 1957, The origin of optical activity, in *Modern Ideas on Spontaneous Generation, Ann. N. Y. Acad. Sci.* **69:**352-368.

Wald, G., 1959, The photoreceptor process in vision, in *Handbook of Physiology* (Amer. Physiol. Soc., J. Field, ed.), Sect. 1, Vol. I, p. 671. Williams and Wilkins, Baltimore.

Wald, G., 1964*a,* The origins of life, Nat. Acad. Sci. U.S.A. Proc. **52:**595.

Wald, G., 1964*b,* General discussion of retinal structure in relation to the visual process, in *Structure of the Eye,* G. K. Smelser, ed., Academic Press, New York, pp. 101-115.

Wald, G., 1970, Vision and the mansions of life, The First Feodor Lynen Lecture in *Miami Winter Symposia,* North-Holland Publ., Amsterdam, pp. 1-32.

Wald, G., and P. K. Brown, 1965, Human color vision and color blindness, *Cold Spring Harbor Symp. Quant, Biol.* **30:**345.

Wald, G., and S. Rayport, 1977, Visions in annelid worms, *Science* **196:**1434-1439.

Walls, G. L., 1942, *The Vertebrate Eye,* Cranbrook Inst. Sci., Bloomfield Hills, Mich.

Warburg, O., and W. Christian, 1938*a,* Coenzyme of the d-amino acid deaminase, *Biochem. Z.* **295:**261.

Warburg, O., and W. Christian, 1938*b.* Coenzyme of the d-alanine-dehydrogenase, Nat.wiss. **26:**235.

Warburg, O., and W. Christian, 1938*c,* Coenzyme of d-alanine oxidase, *Biochem. Z.* **296:**294.

Watanabe, K., 1959, Ultraviolet absorption process in the upper atmosphere, Adv. Geophys. **5:**153.

Waterman, T. H., H. R. Fernandez, and T. H. Goldsmith, 1969, Dichroism of photosensitive pigment in rhabdoms of the crayfish *Orconectes, J. Gen. Physiol.* **54:**415.

Watson, J. D., 1965, *Molecular Biology of the Gene,* Benjamin, New York.

Weiderhold, M. L., E. F. MacNichol, Jr., and A. L. Bell, 1973, Photoreceptor spike response in the hardshell clam, *Mercenaria mercenaria, J. Gen. Physiol.* **61:**24-55.

Wetterberg, L., E. Geller, and A. Yuwiler, 1970, Harderian gland: An extraretinal photoreceptor influencing pineal gland in neonatal rats, *Science* **170**:194-196.

Wetterberg, L., A. Yuwiler, R. Ulrich, E. Geller, and R. Wallace, 1970, Harderian gland: Influence on pineal hydroxyindole-O-methyltransferase activity in neonatal rats, *Science* **171**:194-196.

Wilkins, L. A., and J. J. Wolken, 1981, Electroretinograms from *Odontosyllis enopla* (Polychaete: Sylladae): Initial observations on the visual system of the bioluminescent fireworm of Bermuda, *Mar. Behav. Physiol.* **8**:55-66.

Willstäter, R., and A. Stoll, 1913, *Untersuchungen über Chlorophyll. Methoden und Ergebnisse,* Springer-Verlag, Berlin.

Witman, C. G., K. Carlson, J. Berliner, and J. L. Rosenbaum, 1972, *Chlamydomonas* flagella I, *J. Cell Biol.* **54**:507.

Witman, C. G., K. Carlson, J. L. Rosenbaum, 1972, *Chlamydomonas* flagella II, *J. Cell Biol.* **54**:540.

Wolken, J. J., 1958, The chloroplast structure, pigments and pigment-protein complex, *Brookhaven Symp. Biol.* **11**:87.

Wolken, J. J., 1961*a,* The photoreceptor structure, *Int. Rev. Cytol.* **11**:195-218.

Wolken, J. J., 1961*b,* A structural model for a retinal rod, in *The Structure of the Eye,* G. K. Smelser, ed., Academic Press, New York, pp. 173-192.

Wolken, J. J., 1966*a,* Lipids and the molecular structure of photoreceptors, in *Proceedings of the Symposium on Behavior of Lipids at Interfaces and in Biological Membranes,* F. A. Kummerow, ed., Analabs, Hamden, Conn., pp. 271-274.

Wolken, J. J., 1966*b, Vision: Biochemistry and Biophysics of the Retina,* Thomas, Springfield, Ill.

Wolken, J. J., 1967, *Euglena: An Experimental Organism for Biochemical and Biophysical Studies,* 2nd ed., Appleton, New York.

Wolken, J. J., 1969, Microspectrophotometry and the photoreceptor of *Phycomyces, Int. J. Cell Biol.* **43**:354.

Wolken, J. J., 1971, *Invertebrate Photoreceptors: A Comparative Analysis,* Academic Press, New York.

Wolken, J. J., 1972, *Phycomyces.* A model photosensory cell, *Int. J. Neurosci.* **3**:135.

Wolken, J. J., 1973, Comparative visual systems in microorganisms, plants and animals, tables, in *Biology Data Book,* P. L. Altman and D. S. Dittmer, eds., vol. 2, FASEB, Bethesda, Md., pp. 1269-1270.

Wolken, J. J., 1975, *Photoprocesses, Photoreceptors, and Evolution,* Academic Press, New York.

Wolken, J. J., 1977, *Euglena,* the photoreceptor system for phototaxis, *J. Protozool.* **24**:518-522.

Wolken, J. J., 1984, Self-organizing molecular systems, in *Molecular Evolution and Photobiology,* K. Matsuno, K. Dose, K. Harada, and D. L. Rohlfing, eds., Plenum, New York, pp. 137-162.

Wolken, J. J., and R. G. Florida, 1984, The eye structure of the bioluminescent fireworm of Bermuda, *Odontosyllis enopla, Biol. Bull.* **166**:260-268.

Wolken, J. J., and G. J. Gallik, 1965, The compound eye of the crustacean: *Leptodora kindtii, J. Cell Biol.* **26**:968-973.

Wolken, J. J., and M. A. Mogus, 1979, Extra-ocular photosensitivity, *Photochem. Photobiol.* **29**:189-196.

Wolken, J. J., and M. A. Mogus, 1981, Extraocular Photoreception, in *Photochemistry and Photological Reviews,* vol. 6, K. Smith ed., Plenum, New York, pp. 181-199.

Wolken, J. J., and C. S. Nakagawa, 1973, Rhodopsin formed from bacterial retinal and cattle opsin, *Biochem. Biophys. Res.* Comm. **54:**1262.

Woodward, R. B., 1961, Total synthesis of chlorophyll, *Pure Appl. Chem.* **2:**383.

Wurtman, R. J., 1975, The effects of light on the human body, *Sci. Am.* **233:**68-77.

Wurtman, R. J., and J. Axelrod, 1965, The pineal gland, *Sci. Am.* **213:**50-60.

Wurtman, R. J., J. Axelrod, and D. E. Kelly, 1968, *The Pineal,* Academic Press, New York.

Wurtman, R. J., M. J. Baum, and J. T. Potts, Jr., eds., 1985, *The Medical and Biological Effects of Light,* vol. 453, New York Acad. Sci. Ann.

Yoshida, M., 1979, Extraocular photoreception, in *Handbook of Sensory Physiology,* vol. 7/6A, *Vision in Invertebrates,* A: *Invertebrate Photoreceptors,* H. Autrum, ed., Springer-Verlag, Berlin, pp. 581-640.

Young, J. Z., 1964, *A Model of the Brain,* Oxford University Press, New York.

Young, J. Z., 1971, *Introduction to the Study of Man,* Oxford University Press, New York.

Young, T., 1802, The theory of light and colors, *R. Soc. London Phil. Trans.* **92:**12.

Young, T., 1807, *Lectures on Natural Philosophy,* vol. 1, W. Savage, London, pp. 315 and 613.

Zechmeister, L., 1962, *Carotenoids: Cis-Trans Isomeric Carotenoids, Vitamin A, and Arylpolyenes,* Academic Press, New York.

Zrenner, C., 1985, Theories of pineal function from classical antiquities to 1900, in *Pineal Research Reviews,* vol. 3, R. J. Reited, ed., Alan R. Liss, Inc., New York, pp. 1-40.

Dr. Jerome J. Wolken is a professor of biological sciences at Carnegie-Mellon University, Research Fellow at Mellon Institute, and Research Associate of the Carnegie Museum of Natural History in Pittsburgh, Pennsylvania. Dr. Wolken has been a life-long resident of Pittsburgh, and obtained the B.S., M.S. and Ph.D. at the University of Pittsburgh. He did post-doctoral research at Princeton University and the Rockefeller University. He has been a Visiting Professor of biophysics at the Pennsylvania State University and Visiting Research Professor at the Atomic Energy Commission and the Pasteur Institute, Paris, France; University College, London; the Center for Theoretical Studies, University of Miami; and biology, Princeton University.

Dr. Wolken is a Fellow of the American Association for the Advancement of Science, the Optical Society of America, the American Institute of Chemistry and the Explorers Club, Honoraries of Sigma Xi and Phi Sigma, as well as a member of numerous other scientific societies. He has held research fellowships from the National Institute of Health (NIH), National Science Foundation (NSF), American Cancer Society (NCR), American Philosophical Society and National Council to Combat Blindness. He has worked at various marine biological laboratories throughout the world and has gone on several exploring expeditions in search of how animals see in the deep sea.

Dr. Wolken's major research interests are in photosynthesis, vision, and all photobiological phenomena. He has written eight books on these subjects and 118 research papers in various journals and is the inventor of an optical device for the visually handicapped.